FREE RADICALS
as Studied by
Electron Spin Resonance

By the same author

Spectroscopy at Radio and Microwave Frequencies

FREE RADICALS

as Studied by
Electron Spin Resonance

D. J. E. INGRAM

M.A. (Oxon), D. Phil. (Oxon)

Professor of Physics
University College of North Staffordshire, Keele
(Late of University of Southampton)

LONDON
BUTTERWORTHS SCIENTIFIC PUBLICATIONS
1958

BUTTERWORTHS PUBLICATIONS LTD.
88 KINGSWAY, LONDON, W.C.2

AFRICA: BUTTERWORTH & CO. (AFRICA) LTD.
DURBAN: 33/35 Beach Grove

AUSTRALIA: BUTTERWORTH & CO. (AUSTRALIA) LTD.
SYDNEY: 8 O'Connell Street
MELBOURNE: 430 Bourke Street
BRISBANE: 240 Queen Street

CANADA: BUTTERWORTH & CO. (CANADA) LTD.
TORONTO: 1367 Danforth Avenue

NEW ZEALAND: BUTTERWORTH & CO. (AUSTRALIA) LTD.
WELLINGTON: 49/51 Ballance Street
AUCKLAND: 35 High Street

U.S.A. Edition published by
ACADEMIC PRESS INC., PUBLISHERS
111, FIFTH AVENUE
NEW YORK 3, NEW YORK

First Impression 1958
Reprinted   1960

Set in the Monotype Baskerville Series
Printed in Great Britain by
R. J. Acford Ltd., Chichester

PREFACE

INTEREST in free radicals has grown very considerably during the last few years, both with regard to their role in various practical applications and because of their part in fundamental reactions of physical, chemical or biological importance. It is therefore very fortunate that the technique of electron spin resonance has also been developed at the same time, and it can now be applied as a powerful new method in free radical investigations.

The first measurements on free radicals by electron resonance were made about ten years ago, but during the last two or three years a large number of research groups have started work along these lines, and it is a technique which is still being introduced in various new fields of investigation. From this point of view it might seem premature to write a book on such a subject, since considerable further advances will undoubtedly take place during the next few years. On the other hand the basic ideas, theory and experimental techniques are now well established, and these will remain fundamentally unchanged in future work. Moreover, it would appear that an increasing number of scientists, physical and bio-chemists in particular, are beginning to realize the possibilities of this new technique, and would appreciate an introductory text explaining the basic theory, experimental methods, advantages and limitations, of electron resonance as applied to free radical investigations. It is for such research workers that this book is mainly intended, and it is also hoped that it will serve as a successful introduction for those who are interested in the possibilities of this new field of research, but have previously had no acquaintance with it. At the same time an attempt has been made to summarize all the relevant work published to date, so that a clear idea should be given of the subject as it stands at the moment.

The types of experiment discussed in this book reach into widely different branches of research—into electronics and physics for the basic techniques and design data, into physical, organic and bio-chemistry for most of the actual compounds studied, and now into the biological and medical fields for several of the most interesting applications. It has therefore been somewhat difficult to know at what level the book should be written. An approach via the experimental side has in fact been adopted, and the various methods of free radical investigation are first outlined. The general design considerations of electron resonance spectrometers and the experimental techniques that are used are then considered in some detail. In this way, the book can be divided into two halves by its method of treatment. The first four chapters, which deal with well established design procedure and basic molecular theory, have been written in more detail than the others, since their content should remain unaffected by future work. As an example topics such as ' ultimate sensitivity ' have been considered at some length since this will always be one of the most important features in free radical studies.

The last five chapters, on the other hand, summarize work on different types of free radical system, and a more tentative treatment is given since the interpretation of some of the subjects considered may be modified by future work. These remarks apply especially to the free radicals produced by irradiation. This is a very fascinating field of study and one which is developing rapidly at the moment, but until more detailed background information is built up, some of the interpretations must not be taken as conclusive. The methods of hyperfine structure analysis and the like are described in detail, however, and these spectra afford a very good example of the detailed information that can be obtained by such treatment. It is hoped, in this way, that the book will serve to stimulate further work in the various subjects that are mentioned, and will thus act as an incentive, rather than a summary.

I would like to express my thanks to all those research workers with whom I have had many stimulating conversations on the use of electron resonance in free radical studies, and to Dr. E. E. Schneider, Dr. M. C. R. Symons, and Dr. D. H. Whiffen in particular. I am also very much indebted to my research students for careful checking of proofs, and especially to D. E. G. Austen. I would also like to acknowledge the support and interest shown in our own work at Southampton by Professor E. E. Zepler.

SOUTHAMPTON
July, 1958

CONTENTS

vii

THE DETECTION AND PROPERTIES OF FREE RADICALS

1.1 INTRODUCTION

THE study of free radicals is now reaching into ever-widening fields of research, ranging from solid state physics through chemistry to the biological and medical sciences. The kinds of system that are being investigated vary from the detailed study of electron spin-spin and exchange interaction, as observed in such stable crystals as diphenylpicrylhydrazyl, to the transient and highly unstable radicals associated with photochemical decomposition, polymerization, enzyme reactions and the like. Much attention is therefore being focused on the different methods of detecting and investigating free-radical systems, and during the last few years the advent of electron resonance has provided a powerful new tool for this work.

Several standard physical and chemical methods have been employed in free-radical studies for some time, and these are briefly summarized in a later section of this chapter, but the microwave methods of electron resonance have two outstanding advantages and most of the book is concerned with these techniques. The two features are, first that the sensitivity of detection in the condensed phase is far greater than that of the older methods, since less than 10^{12} unpaired electrons per gramme, or less than 10^{-9} molar concentration of radicals, can be observed. Secondly, very detailed information on the actual nature of the radicals, and the molecular orbitals occupied by the unpaired electrons, can be obtained from the highly resolved hyperfine structure which is normally observed in the spectrum. These features not only allow identification of different radical species present in a reacting system, but also allow the rate of reactions to be followed and a study of dynamic concentrations to be made.

In presenting the study of free radicals by electron spin resonance an approach via the experimental technique has been adopted. Thus this first chapter begins with an outline of the properties of free radicals and a summary of the standard methods of investigation. A brief outline of the theory and technique of electron resonance and other microwave methods is then given, together with some of the practical considerations involved in their study.

The second and third chapters then deal more fully with the experimental methods, so that those wishing to start work in this field can set up the necessary apparatus. The ensuing chapters are then devoted to a discussion of the different types of radical systems that can be studied, and the particular points associated with each system are noted. The investigation of stable radicals is considered first, and then the following chapters describe those formed by irradiation, thermal and chemical means, and during such processes as polymerization. The application of these studies to biological and medical problems is finally discussed in some detail.

1.2 Free Radicals and Unpaired Electrons

One of the most general definitions that can be given for a free radical is that it is ' a molecule, or part of a molecule, in which the normal chemical binding has been modified so that an unpaired electron is left associated with the system '. In this way all the paramagnetic salts of transition group elements are eliminated as they have unpaired electrons associated with their *normal* chemical bonding, and no modification is required to produce them.

The definition includes, however, all the organic radicals which are formed by abstraction of a hydrogen atom from a ring or hydrocarbon chain, and in fact any system in which the unpaired electron is moving in a molecular rather than in an atomic orbital. There are, of course, some borderline cases such as the inorganic compounds which contain unpaired electrons in their normal state of chemical binding, e.g. O_2, NO and ClO_2. Since the normal chemical binding has not been modified in these cases they do not literally come under the above definition of a free radical, and in the literature are generally not treated as such. Their unpaired electrons are associated with molecular orbitals, however, and they have several features in common with other radical systems, and their study is therefore included in this book for the sake of completeness.

In a similar way the defects and irradiation damage produced in ionic crystals form a borderline case. The unpaired electrons present in these are often associated with localized atomic, rather than molecular, orbitals, but since they are only produced by a modification in the normal chemical binding they do fit in with the above definition, and are included under the general heading of ' Radicals produced by Irradiation '.

It is evident from these considerations that the existence of an unpaired electron in the outer orbitals of the system is an essential feature of a free radical. The presence of such an electron endows the system with two very characteristic features; first an extremely high reactivity in chemical actions, and, secondly, a magnetic moment, due to the uncompensated spin motion of the odd electron. It therefore follows that there are, in general, two kinds of experiment that can be employed to detect and study free radicals. The first is by the use of chemical methods, based on their high reactivity, and the second is by the use of physical methods, based on their magnetic properties. In certain cases the high chemical reactivity can be masked by steric hindrance or trapping effects, and stable or temporarily stable radicals are produced. Such cannot be easily studied by methods based on high chemical reactivity, whereas the physical methods are just as applicable since the magnetic moment of the unpaired electron remains unchanged. For this reason, and also because of its accurate quantitative possibilities, the magnetic resonance method may claim to be more fundamental and reliable.

If the time-scale of the measurements can be suitably reduced then standard spectroscopic methods can be used to identify radical species, and a very good example of this is the recent work on flash-photolysis. It may be noted, however, that the main application of this type of technique is to studies of radicals in the gaseous state, whereas magnetic resonance investigations are mainly carried out on the liquid and solid phases; and these two new techniques are therefore complementary rather than competitive.

Most of the experiments and results described in this book are therefore based on the fact that every free radical has a magnetic moment associated with its unpaired electron. By employing suitable methods of detection the concentration of these magnetic moments can be measured down to very low values, and their interactions with the various atoms of the radical can be studied. The possibility of investigating free radical processes by standard magnetic susceptibility measurements was realized and exploited at quite an early stage[1], but it was not until the advent of electron spin resonance that the real power of these magnetic methods became apparent.

1.3   PROPERTIES UNIQUE TO RADICALS

If small concentrations of free radicals are to be detected in a large mass of other material, use must be made of some property that is unique to the free radical. It has been seen that such properties can be of a physical or chemical nature and these are now considered in more detail.

3

The high chemical reactivity was in fact the most convincing experimental proof that a free radical had been obtained when GOMBERG[2] first produced triphenylmethyl. It was found that the parent hexaphenylethane could only be prepared in the complete absence of air and that it would react instantly with oxygen, iodine and nitric oxide to give derivatives of triphenylmethyl. These two features of high abnormal chemical reactivity, and identification of the end-product as the result of a radical action, have remained as two of the standard chemical methods of free radical detection.

A more direct and quantitative way of detecting high chemical activity was employed by PANETH[3] and his co-workers, in the form of deposited metal mirrors. Free radicals passing across the surface of these will react with the metallic film and remove the mirror, and a very direct confirmation of their presence is thus obtained. The nature of the radicals can again be established in these cases by analysing the end-products that are formed, which may be collected in liquid-air traps if necessary.

Although the mirror technique can be widely applied to radicals in the gaseous state, it is not very applicable to those formed in the condensed phase. In these cases some other methods of detecting high or abnormal chemical reactivity must be employed, and the formation of unexpected end-products is again the most direct method. An example of this is the chain reactions which occur during polymerization in the presence of a radical catalyst. This property was used in the early work of FREY[4] and SICKMAN and ALLEN[5] who showed that butane plus 1 per cent dimethyl mercury, or acetaldehyde plus 5 per cent azomethane would decompose and form chain systems at much lower temperatures than either of the pure compounds would by themselves. Such work as this has developed enormously over recent years, but the general principle has remained that high chemical reactivity, as shown by end-products produced under unusually mild conditions, can be taken as strong evidence for radical formation and reaction.

The detection of radicals by chemical means can also be placed on a quantitative basis by kinetic studies of the reactions, and accurate values of activation energies obtained in this way. In a similar way polarographic techniques can also be employed in which current–voltage curves, measured during reduction or oxidation, can give evidence that one electron changes are taking place. The existence of odd-electron states is also often associated with a pronounced colour in the system (as in the case of paramagnetic transition group salts), and this allows colorimetric methods to be used to estimate radical concentrations.

4

The physical methods that are employed to detect and measure free radical concentrations are nearly all based on the fact that the unpaired electron will have a magnetic moment. The simplest way in which the magnetic moments of the unpaired electrons can be detected is by a direct measurement of the susceptibility of the sample. The presence of the radicals will then give a paramagnetic component, and a quantitative measurement of this can be used to estimate the number of radical centres present. This method was in fact used in 1935 by MULLER[6] to measure the values for the dissociation of hexaphenylethane in benzene solutions at different temperatures. Susceptibility measurements have been employed in an increasing number of cases since this time, and some of the different types of apparatus, including the standard static methods and the more recent dynamic systems, are discussed in detail in the next chapter together with the fundamental theory of para-magnetic susceptibility. There are two important sources of error that may enter into the susceptibility measurements, however. One is that any small amount of ferromagnetic impurity in the sample will give rise to very erroneous readings due to its much larger susceptibility. The other is that correction must be made for the inherent diamagnetism of the compounds, and this may be quite appreciable, and difficult to estimate accurately, for the larger aromatic systems. Magnetic resonance methods, however, are devoid of both of these errors since ferromagnetic impurities will give an absorption well removed from the free-spin value, and the diamagnetism is not measured at all by the resonance technique. The theory and experimental methods of electron resonance are considered later in this chapter and in more detail in Chapter 2. Here it may just be noted that it is essentially a dynamic method, in which the energy associated with the spin and moment of the unpaired electron is measured. The unpaired electrons of the free radicals are thus studied quite apart from any other property of the rest of the sample, and owing to the very high resolution available at these frequencies, detailed information can often be obtained about the structure of the radicals, as well as quantitative measurements made on their concentration.

1.4 STANDARD METHODS OF DETECTION

A brief summary is now given of some of the well established methods of free radical detection; this is not intended as any authoritative account but to serve as a comparison for the resonance techniques.

1.4.1   *The Mirror Technique*

This technique was first used by PANETH and HOFEDITZ[3] in 1929 to demonstrate the existence of the free methyl radical, and has been applied since then to a large number of other gaseous systems. The apparatus used in their initial work is shown in *Figure 1*. A rapid

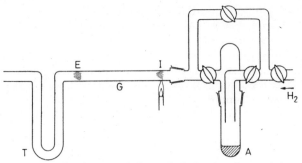

Figure 1.   The detection of free radicals by the mirror technique.
A, *Tube containing lead tetramethyl*;   T, *Liquid air trap*;   E, *Position of first deposit*;   I, *Position of second deposit*

current of hydrogen or nitrogen, at a few mm pressure, was passed over lead tetramethyl at 193° K held in the tube, *A*, and the resulting mixture of hydrogen and lead tetramethyl vapour was then passed down a long glass tube, *G*, ending in a liquid air trap, *T*. The tube was first heated locally near the exit, at *E*, and a thin layer of metallic lead was deposited on the glass walls as the lead tetramethyl decomposed. The tube was then heated near the inlet, at *I*, when another lead mirror was formed, but at the same time it was noticed that the original mirror at *E* disappeared. Moreover the time taken for this mirror to disappear depended on the distance between it and the second point of heating at *I* and on the rate of gas flow down the tube. It was thus evident that some entity, formed during the decomposition of the lead tetramethyl at *I*, was being swept down the tube and was then reacting with the cold deposit of metallic lead, which constituted the mirror at *E*. It was easily shown experimentally that none of the normal stable gaseous decomposition products would react with the mirror, and the occurrence of highly reactive methyl radicals was therefore postulated as present in the gas stream. This was further confirmed by the fact that the same action would also occur with other deposited mirrors, such as zinc or antimony, and that if the final end products were caught in the liquid air trap and analysed they were found to be zinc or antimony trimethyl.

6

This mirror technique has been widely adapted since its initial use, and quite a large number of other metals have been employed. The essential feature is, of course, the formation of a metallo-organic salt from the cold metal surface, and the identification of such a resultant reaction product can be taken as definite evidence of radical formation, even when the action on the mirrors themselves cannot be observed. Another more recent development has been the inclusion of radioactive tracers in the mirror surface, and the transference of these to the liquid air trap can be made a very sensitive test for the presence of small free radical concentrations.

The selective action[7] of radicals on mirrors of different metals, and quantitative rate studies on the disappearance of the mirrors under different conditions[8], have rendered this technique a very powerful method for the study of short-lived gaseous radicals.

1.4.2 Detection via Abnormal Chemical Reactivity

The abnormally high chemical reactivity associated with free radicals is really the basis of most chemical means used for their identification. The mirror technique, described above, is in fact just one example of a possible method of detecting such high reactivity. Although this technique can be used in most gaseous radical studies it is not very applicable when solids or liquids are being studied and in such cases other means of detecting the abnormal reactivity must be employed. It has already been seen that there are in fact two basic methods of identification, either by analysis of the end-products of the reaction, or by measurement of the rates of the reaction. In the first case the presence of free radicals is shown by the abnormal products which are formed, i.e. they are detected by the fact that they have initiated otherwise forbidden chemical reactions. In the second case the values of the rate constants will show that entities with very high combining energy have been present at one stage of the reaction process. A very large amount of work has now been carried out, employing both of these principles, and no attempt will be made to review it here, but one or two examples may serve to illustrate the general features.

The initiation of otherwise forbidden chemical reactions has been used throughout the whole field of polymer chemistry to demonstrate the presence of radical reactions. The original experiments of Frey[4] which showed that butane decomposed at much lower temperatures than usual, if 1 per cent of dimethyl mercury were added, is a good example of the kind of reasoning involved. As another example, some of the most direct evidence for the existence of radicals comes from the presence of chain reactions, and their

7

effect on the rate constants. A chemical reaction involving free radicals can be divided into three phases insofar as its rate characteristics are concerned. There is an initial period during which the free radicals are formed and the reaction rate accelerates with time until the rate of radical generation is equal to their rate of destruction or recombination. There is then a period of equilibrium during which the reaction velocity remains constant and independent of the bulk concentrations of the main reactants. This is followed by a decay as the supply of one of the reactants becomes exhausted. If a reaction takes place with this type of variation in its rate constant, with an intermediate region in which the velocity is independent of the bulk concentrations of the main components, it can be concluded that a form of chain reaction is taking place.

An example of a very direct method of measuring high chemical reactivity is the use of platinum wires to detect the recombination of gaseous free atoms or radicals[9]. If the recombination of the radical species takes place on the metal surface, considerable energy will be involved and the temperature of the platinum wire will therefore rise. If it is made the fourth arm of a bridge it is possible to adjust for balance quite precisely. A measure of the decrease in wattage supplied to the platinum detector for constancy of temperature can thus be made, and accurate values for the recombination energies determined. This is only one example of the many techniques employed to detect high chemical activity and determine recombination energies, however, and for comprehensive surveys of this field several standard books can be consulted[10,11].

1.4.3 *Magnetic Susceptibility Measurements*

It has already been seen that the existence of a magnetic moment, associated with the unpaired electrons, gives the free radicals an inherent paramagnetic susceptibility, and that this can therefore be used to detect their presence and measure their concentration in a very precise manner. The general theory of paramagnetic susceptibility and the more modern experimental techniques of dynamic measurement are considered at some length in the next chapter. As a comparison with the other methods, however, one of the earlier standard types of apparatus, as used for the investigation of free radical changes taking place during a slow chemical reaction[12], is shown in *Figure 2*. This is based on the Gouy and Quincke methods of susceptibility determination, and uses the fact that, if one end of a sample with susceptibility χ is in a field of value H_0 and the other end is in zero field, then there is a resultant force on the specimen equal to $\frac{1}{2}.\chi.a.H^2$; where a is the cross-sectional area of the specimen. In this particular apparatus the sample is in liquid form, and the force due to the magnetic field is balanced by the hydrostatic pressure produced by a change in the level of the liquid in the tube. The force acting on the specimen, and hence its susceptibility, can thus be directly calculated from this change in height of the liquid level when the magnetic field is applied. The

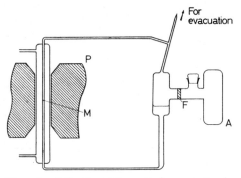

Figure 2. Estimation of free radical concentration by measurement of susceptibility. A, Reaction tube; F, Sintered glass filter; P, Magnet pole pieces; M, Liquid meniscus level observed by travelling microscope (After Roy and Marvel[12])

free radical reaction is carried out in the sealed and evacuated bulb, *A*, after the reagents have been inserted via the glass stopper, and when the reaction has progressed for the required time the products are filtered through the sintered glass filter, *F*, into the U-tube section. The susceptibility is then measured by inserting the narrow arm of the U tube into the magnet gap. Various other methods of susceptibility measurement, including the very sensitive rotation techniques, are summarized in the first section of the next chapter.

Another somewhat indirect method of detecting the presence of free radicals via their magnetic properties is possible by utilizing their catalytic action on the ortho-para-hydrogen conversion. The interconversion of the ortho (parallel nuclear spins) and para (opposed nuclear spins) forms of the hydrogen molecule does not take place quickly under normal conditions of gaseous collision, and it is possible to keep pure para-hydrogen for weeks at room temperature before it reaches the equilibrium mixture of one part para to three parts ortho-hydrogen. This interconversion can be greatly accelerated, however, if a third body, possessing a magnetic moment, is added to the mixture. It can be shown[13] that collisions with such molecules will produce such a large magnetic perturbation in the hydrogen molecule that relaxation between the ortho- and para- states is greatly increased. It therefore follows that the presence of a free radical with its unpaired electron can be detected by measuring the rates of interconversion of ortho- and para-hydrogen. This method can be applied to gas-phase studies such as detection of nitric oxide, or to liquid-phase studies such as the estimation of triphenylmethyl[14], or to the action over solid surfaces such as 1,1-diphenyl-2-picryl-hydrazyl[15] or charcoals.

1.4.4 *Investigations by Mass Spectrometry*

Mass spectrometry is being applied to a growing number of free radical problems and, in certain cases, can give very accurate measurements on the nature and concentrations of the radical species. The possibilities of this technique in studying the reactions and properties of free radicals was first demonstrated by ELTENTON[16], who showed that radicals produced in thermal decomposition reactions could be detected in the presence of large amounts of other end-products. Later work[17] showed that quantitative measurement of the ionization potentials of the radicals could also be determined from electron impact experiments, as well as detailed rate studies carried out in suitable systems.

The general principle of the method is that the radicals and other reaction products are allowed to enter the ionization chamber of the mass spectrometer and are there ionized by electron bombardment. The mass of the resulting ions is then determined by the spectrometer measurements, and its identity as a radical established from the detailed atomic constitution of the molecule. Very high resolution in mass measurement is now available and an unambiguous determination of the numbers of the different atoms in the molecule or radical is usually possible. If this method is to be used to demonstrate that radicals are present in the initial reaction mixture, care must be taken to ensure that the same radicals are not formed from stable molecules by the electron beam bombardment. Fortunately this is not a very difficult problem as much more energy is usually required to break a molecular bond and form an ionized radical than to ionize a radical which already exists. The energy of the electron beam is therefore kept to within a few electron-volts of the ionization potential of the radicals concerned, and creation of new radicals is then very unlikely.

The method has, of course, some inherent limitations. It can only be applied to cases in which the radicals can be made to enter the ionization chamber, and the permissible pressure range is also somewhat restricted. It has been applied with considerable success

10

to such systems as combustion products, electrical discharges and reactions with excited atoms. Radicals varying from hydroxyl and methyl groups to benzyl and phenoxy groups have been detected by this technique. Further details of the applications of this method can be found in various review articles[18,19].

1.4.5 *Absorption Spectroscopy and Flash Photolysis*

The study of radicals by the normal methods of absorption spectroscopy in the ultra-violet, optical, or infra-red region has become of increasing importance in recent years. The main experimental problem has been either to artificially lengthen their lifetime by trapping or similar means, so that their absorption spectra can be recorded by normal methods; or to develop high-speed techniques so that their spectra can be studied in the first few milliseconds after formation.

Radical concentrations in dynamic equilibrium can be relatively easily investigated by normal methods, and the identification of the a-bands of ammonia, produced in gas discharges or oxy-ammonia flames, as due to NH_2 radicals, is a classic example of this type of work. A new method of studying absorption spectra of radicals which were not in dynamic equilibrium was introduced by LEWIS and LIPKIN[20], however, in which the parent molecule was irradiated in a rigid glass at low temperatures. They observed the absorption spectra of triphenylmethyl and other large aromatic radicals by this means, and the work has been extended to other cases by NORMAN and PORTER[21]. This technique of trapping radicals in a medium which is sufficiently mobile not to inhibit their initial formation, but sufficiently rigid to trap them once they have been produced, is being increasingly used in free radical studies, and can be employed with considerable success in electron-resonance work[22].

Another approach to the study of unstable radicals by absorption spectroscopy has been the development of the flash photolysis technique by NORRISH and PORTER[23], HERZBERG and RAMSAY[24], and DAVIDSON et al.[25]. The essential principle of this method is to employ two very powerful flashes; the first produces the radicals by photochemical decomposition, and the second, which follows microseconds or milliseconds later, is used to record the absorption spectra of the radicals then in existence. The energy required to initiate these flashes is of the order of one thousand joules, and condensers of about $100\mu F$ capacitance need to be charged to between four and eight thousand volts. A typical apparatus, as used in the earlier work, consisted of the initial discharge lamp, which had a quartz envelope about 50 cm long with the discharge

11

taking place through an inert gas such as argon or krypton. The flash was of the order of 10^{-4} seconds duration, and although the photochemical efficiency may only be about 10 per cent, a very large decomposition of suitable photochemically reactive species could be produced. The reaction vessel in which the radicals were formed was also of quartz and of the same length as the discharge lamp, and the two were placed close together and surrounded by a reflector, which concentrated the light energy on to the reaction tube. The absorption spectra of the radicals so formed was then recorded by striking another flash in a second discharge tube. This flash need not be of such high energy as the first, but should produce a continuous spectrum. The initiation of both flashes could be controlled by triggering electrodes, and the time interval between them could be varied from $30\mu\text{sec}$ upwards by the use of suitable control circuits. Comparison of the absorption spectra taken at different time intervals thus enabled information on the rates of the radical reactions to be obtained.

This technique is a very powerful method for studying gaseous free radicals and gives detailed information on both their structure and kinetics. It is limited to gaseous systems, however, and only those in which the reaction can be photochemically induced or initiated. In both the flash photolysis and the low temperature trapping experiments, the identification and structural analysis of the radical species proceeds from the absorption spectra in the normal way. Thus an analysis of the rotational and vibrational bands not only allows the actual constitution of the absorbing entity to be determined, but also its bond lengths and angles.

1.4.6 Other Methods

The most important of the standard methods of free radical detection have been briefly summarized above, but this is not a comprehensive list, and other techniques have also been employed such as those based on polarographic[26] or colorimetric methods, or on the employment of other free radicals such as 1, 1-diphenyl-2-picryl-hydrazyl or iodine vapour as scavenging agents. New methods of kinetic study have also been devised, such as those employing shock waves[27] and other similar techniques will no doubt soon be applied to aid in these investigations.

This brief capitulation of the principles behind the usual methods of free radical study may serve as a useful comparison for the electron resonance techniques with which this book is mainly concerned. It might be noted in this connection that the systems most readily studied by the standard methods are those involving free radicals in the gaseous phase. Mass spectrometric or flash photolysis methods can be easily applied to these and yield accurate quantitative data. Study of radicals in the liquid or solid phase is not open to such direct measurement by these methods, however, and the more indirect chemical analysis of end products or reaction kinetics must normally be employed. It is just in these cases how-ever, that the electron resonance methods can be applied to greatest advantage and hence such techniques as flash photolysis and electron resonance can be regarded as complementary methods for free radical study, the former being applied mainly to gaseous reactions and the latter, at the moment, more to liquid or solid state systems.

1.5   MICROWAVE METHODS OF DETECTION

Microwave spectroscopy can be divided into two main branches; first, experiments which measure the molecular resonant frequencies in the absence of any large perturbing fields and secondly, those in which the energy level system is considerably altered by the appli-cation of a large magnetic field, and resonance absorption is then obtained from transitions between the new levels.

The first type of experiment is essentially the same as in infra-red spectroscopy, except that the wavelengths are now much longer than normal and most of the molecular frequencies are associated with simple rotational motion. The techniques are very different, of course, but the method of analysing the results is the same as in normal infra-red practice. Bond lengths and angles can be deter-mined with high precision, because frequencies can be measured very accurately and the spectra are extremely well resolved. This extension of absorption spectroscopy from the infra-red into the microwave region has been used to determine a large number of molecular structures and has proved to be a very powerful technique in the analysis of the lighter gaseous molecules[28]. It was therefore hoped that this might also provide a powerful new tool with which to detect and identify gaseous radical species, the new rotational levels of the radical giving spectra which would be easily identified. This initial promise has unfortunately been unfulfilled and very few radicals have in fact been detected by the normal methods of gaseous microwave spectroscopy. There are several reasons for this failure, including

(a) the initial difficulty of introducing the radicals to the absorp-tion cell, or

(b) the heating of the latter to preserve a dynamic concentration, and

(c) the rapid radical recombination that occurs if Stark electrodes or other metal surfaces are placed in the reaction stream.

The only really successful experiment in which unstable radical species have been detected and identified by normal gaseous microwave spectroscopy has been the work of DOUSMANIS, SANDERS and TOWNES[29] on the OH and OD radicals. This work is quite distinct from the electron resonance type of investigation, which is described in detail in the rest of the book, but an account of the experiment is included here to illustrate the accurate data that can be obtained from work at microwave frequencies. A diagram of their apparatus is given in *Figure 3*. The OH or OD radicals were

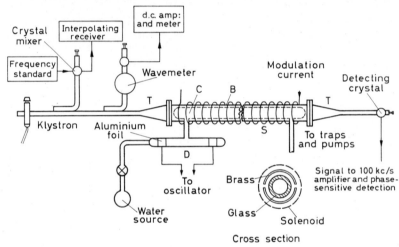

Figure 3. Gaseous microwave spectrometer for free radical study. D, Discharge tube; C, Glass absorption cell; B, Brass tube acting as cylindrical waveguide (After Dousmanis, Sanders and Townes[29])

produced in the discharge tube, *D*, which had external aluminium electrodes fed from a 300 watt oscillator operating at about two megacycles. The products of the discharge were then passed directly into the absorption cell itself, *C*, which was made of low-loss glass, 150 cm long and blown with minimum thickness at the two ends. This glass tube was surrounded by a brass pipe, *B*, split in half along its length and with holes for the inlet and outlet tubes. This pipe acted as a cylindrical waveguide operating in the H_{11} mode, and the microwave power entered and left through tapered sections, *T*, which were connected to standard sizes of rectangular waveguide. Magnetic modulation, rather than the normal Stark modulation[28], was employed in this experiment, for two reasons. First, since the radicals will have an unpaired electron, they will have a large magnetic moment and Zeeman effect, and should thus be readily detected by small magnetic field modulation, whereas all the other reaction products in the absorption cell will be unaffected to a first order, and will therefore produce no spectra. Secondly, Stark modulation requires a metal electrode down the centre of the absorption tube, and this would not only be difficult to insert but would also cause a high rate of radical recombination. The magnetic field Zeeman modulation could be very easily applied, however, by a simple solenoid, S, wound around the brass tube. This produced a field of a few gauss axially down the absorption cell, and was modulated at a frequency of about 100 kc/s. The microwave spectrometer itself was of the normal modulation type[30], with a klystron as the microwave power source, and a silicon crystal as detector followed by a narrow-band amplifier, phase-sensitive detector and pen-recorder.

The energy level system of the free OH radical is very complex since the rotational terms are perturbed to a considerable extent by coupling with the electronic motion and a large Λ-doublet splitting is produced. The absorption lines which were observed in the 7,000 Mc/s to 37,000 Mc/s region were, in fact, due to direct transitions between the Λ-doublet levels of the different rotational states in the ground vibrational level. A hyperfine structure is also obtained on each transition, due to the magnetic moment of the proton or deuteron, and accurate measurements of these splittings can also be used to calculate the unpaired electron distribution in the molecule. The detailed measurements obtained from this experiment are, in fact, a very good example of the wealth of information that can be deduced from precise microwave measurements. Not only were the radicals detected and identified as $O^{16}H$, $O^{18}H$ and $O^{16}D$ groups, but the detailed theory of the $\Pi_{\frac{1}{2}}$ and $\Pi_{\frac{3}{2}}$ states was checked to 1 part in 3,500. The unpaired electron distribution was also calculated from the hyperfine structure separations as

$$(1/r^3)_{Av.} = (0\cdot75\pm0\cdot25)\times10^{24}\text{cm}^{-3} \qquad \ldots\ldots(1.1)$$

$$\text{and } (\sin^2\chi/r_3)_{Av.} = (0\cdot49\pm0\cdot01)\times10^{24}\text{cm}^{-3}$$

Studies on the lifetime and reactivity of the radicals were also made by studying radical concentrations at different points down the absorption cell. This was effected by applying the modulating Zeeman current only to short sections of the cell at a time. The

THE DETECTION AND PROPERTIES OF FREE RADICALS

effect of inserting different surfaces into the radical stream was also investigated by mounting these on a retractable rod placed in a side arm. In this way it was possible to show that surfaces of copper, graphite and nickel greatly decreased the OH radical concentration whereas those of aluminium, teflon and potassium chloride had no appreciable affect.

Two gaseous microwave spectrometers have recently been constructed for the specific purpose of studying free radicals. In both of these the metallic surface in the absorption cell has been reduced to a minimum in order to prevent radical recombination as much as possible. In the spectrometer designed by COSTAIN[34], the absorption cell consists of a pyrex tube with a polystyrene lens and microwave horn at each end. The lenses also serve as vacuum windows and in this way all metallic surfaces are completely eliminated from the cell. HURLE and SUGDEN[35] employed a somewhat similar principle with horn feed to a large diameter pyrex tube, but used a thin wire along the centre of the tube to support a surface wave, instead of a lens system. No detailed reports of actual free radical studies with these spectrometers have yet been made, however.

1.6   ELECTRON SPIN RESONANCE*

The second main branch of microwave spectroscopy is that in which a large magnetic field is employed to alter the normal energy level system so that transitions can take place in the microwave region. This is in fact the method used in electron resonance studies, and which will now be considered in somewhat more detail. The first condition to be fulfilled, if a system is to be investigated by electron resonance techniques, is that it should possess an unpaired electron with its associated magnetic moment. It is for this reason that the technique is so applicable to free radical studies, since this is also just the condition implied in the definition of a free radical. The method of electron resonance is also of very general application, in that the resonance conditions are not determined by any particular resonant frequency of the free radical structure, but by the frequency associated with the energy of the electron moment in the applied magnetic field. It is therefore known beforehand at what frequencies and fields any free radical spectra will occur.

---

* The term ' electron spin resonance ' is used throughout this book to describe the magnetic resonance experiments in which the electron, rather than the nuclear, spin changes orientation. The term ' paramagnetic resonance ' has often been used to describe this kind of work in the past, but since this title does not differentiate clearly between the electron and nuclear cases, it has been decided that the more explicit term of ' electron spin resonance ' or ' electron resonance ' should be used in future.

The case of a simple free radical in which the unpaired electron is not coupled to any nuclei in the radical can be considered first. When the sample is in zero-magnetic field, the spins and magnetic moments of the unpaired electrons will be pointing in random directions and will all have equal energy. If a d.c. magnetic field is now applied across the specimen, however, the electrons will align themselves either with their spins and moments parallel to the applied field, or anti-parallel to it, no intermediate orientations being allowed by the quantum conditions since $S=\frac{1}{2}$. The electrons will therefore now fall into two groups, and these groups will have different energies, since those with their spins aligned parallel to the field will have an energy of $\frac{1}{2}g\beta H$ less than the zero-field value, and those with their spins aligned antiparallel to the field will have an energy of $\frac{1}{2}g\beta H$ greater than the zero field value. The 'spectroscopic splitting factor' or 'g-value' is a measure of the contribution

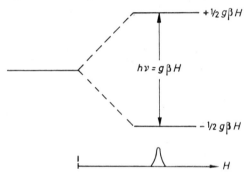

Resonance condition :- $h\nu = g\beta H$

$\therefore \nu = 2 \cdot 8 \times 10^{6}\,H$ for free electron

Figure 4.   The basic electron resonance condition. Divergence of energy levels of single unpaired electron with no hyperfine interaction

of the spin and orbital motion of the electron to its total angular momentum, and has a value of 2·0023 for a completely free spin. (The small divergence from 2·000 comes from the relativity correction.) The Bohr magneton, β, equal to $\dfrac{eh}{4\pi mc}$, serves as a constant to convert the angular momentum to magnetic moment in electromagnetic c.g.s. units.

The state of affairs in the presence of an applied magnetic field can therefore be represented by the energy levels of *Figure 4*. It is seen that the splitting between the two energy levels increases linearly with increasing magnetic field and for a given field, H, is equal to $g\beta H$. If radiation of frequency ν is now fed to the sample, some of the electrons in the lower state will absorb energy and jump to the upper state in the process provided that

$$h\nu = g\beta H \qquad \ldots (1.2)$$

At the same time those in the upper state will be stimulated to emit radiation of frequency ν and fall to the lower state. For any system in thermal equilibrium there will, however, be more electrons in the ground state than in the upper state, and hence there will be a net absorption of the radiation of frequency ν. In most cases the distribution of electrons between the two states is given by the Maxwell–Boltzmann expression in which the ratio of the number in the upper state, n_1, to that in the lower state, n_2, is given by

$$\frac{n_1}{n_2} = e^{-\frac{\Delta E}{kT}} \qquad \ldots (1.3)$$

where ΔE is the separation between the two levels, k is the Boltzmann constant and T is the absolute temperature. As a typical example this gives a ratio of 0·9984 for room temperature and an energy separation corresponding to 3 cm wavelength. This fact also explains why greater sensitivity is obtained by working at low temperatures since T is then reduced, and the difference between n_1 and n_2 increased, so that a larger net absorption occurs.

It also follows that, if this resonance absorption of radiation is to continue, there must be some other mechanism, apart from the stimulated emission, which allows the electrons in the upper state to lose energy and drop back to the lower state. This mechanism must allow energy transfer of $h\nu$ by interaction with some system other than the incident radiation. If this were not so the larger number absorbing energy in the ground state would quickly tend to equalize n_1 and n_2, with the result that no further absorption of radiation would occur. There are various ways in which electrons can lose energy and so fulfil this condition. These are termed relaxation processes and when the energy of the spins is being shared with the thermal vibrations of the solid as a whole the effect is known as a 'spin-lattice interaction' and its strength can be measured by a 'spin-lattice relaxation time'. This is the time in which an initial excess of energy given to the spins will fall to $\frac{1}{2}$ of its value. A strong spin-lattice interaction therefore produces a short relaxation time, and a weak interaction produces a long relaxation time.

It therefore follows that, if a system has a long spin-lattice relaxation time, the thermal equilibrium may be disturbed if too high a power level of the incident radiation is used. This ' saturation ' effect very seldom occurs in electron resonance studies of paramagnetic atoms, since their spin-lattice interactions are strong and the relaxation times therefore very short (of the order of 10^{-8} sec). In free radical studies, however, the spin-lattice relaxation times can be very long (of the order of seconds), and saturation by relatively low microwave powers can occur. This saturation effect is observed experimentally by an apparent decrease in the relative size of the absorption, and usually by an increase in the line width. The different types of saturation that can occur are discussed in detail in Chapter 4, but for the rest of this section it will be assumed that thermal equilibrium is present between the spin levels and the rest of the specimen.

It can be seen by reference to *Figure 4* that, in principle, electron resonance absorption can take place at any frequency, provided the value of the magnetic field is adjusted to satisfy the equation $h\nu = g\beta H$. Electron resonance of some free radicals has in fact been carried out at fields of a few gauss and frequencies in the normal radio wave band[31], but if high sensitivity is required it is best to work at as high a value of magnetic field as possible. This can be seen from equation 1.3, as an increase in H will increase ΔE and increase the difference between n_1 and n_2, thus producing a larger net absorption. There are therefore two ways in which the intensity of absorption, and thus the sensitivity, can be increased

(a) by lowering the temperature, or
(b) by increasing the magnetic field strength and the resonant frequency.

Other things being equal, it can be shown that the intensity of absorption varies as ν^2, and high values of ν and H are therefore essential for high sensitivity. The gain in sensitivity between radiofrequency and microwave frequency measurements is not in fact as great as suggested by this relation because the efficiency of detection falls off at the higher frequencies, but a substantial improvement is nevertheless obtained.

The upper limit of field strength and resonant frequency is set by purely practical considerations. It is necessary to apply a uniform magnetic field over a volume of about one c.c. in most free radical work, and it is hard to design a magnet which will produce fields much in excess of 10,000 gauss and fulfil these conditions. At the

19

same time there are few klystrons or other microwave sources available with much output power at frequencies greater than 40,000 Mc/s, and the upper limit of operation for most electron resonance studies is at 8 mm wavelength ($\nu=36{,}000$ Mc/s, $H=13{,}000$ gauss).

There is also one other practical factor which affects the choice of frequency for optimum sensitivity, and that is the nature of the specimen and its holder. If the specimen is in the form of a small single crystal which can be placed directly into the cavity resonator by itself, then maximum sensitivity will probably be obtained with an 8 mm wavelength spectrometer. If, on the other hand, the specimen is an amorphous powder which must be held in a glass or quartz tube then better sensitivity is likely to be obtained at 3 cm wavelength. This is because it is necessary for the glass to be of a certain minimum thickness for mechanical strength, and whereas this thickness is only a small fraction of the size of a sample that can be placed in a 3 cm wavelength cavity, it will be a very considerable fraction of the size of a sample that can be placed in an 8 mm wavelength cavity.

Only a very small sample volume of this type can therefore be placed in an 8 mm spectrometer, and even so considerable damping due to the glass envelope is likely to occur. The klystrons and other microwave equipment available at 3 cm wavelengths are also more reliable and less prone to frequency drift or mismatching than those at present available at 8 mm or 1·25 cm wavelength (the other standard radar waveband). As a result of these various factors most electron resonance spectrometers which have been built for free radical studies and high sensitivity, have employed wavelengths of about 3 cm ($\nu=9{,}000$ Mc/s) with corresponding magnetic fields of about 3,000 gauss.

The basic requirements for an electron resonance spectrometer can now be listed. These are:

(1) A source of radiation at about 3 cm wavelength.

(2) An absorption cell into which the specimen is placed, and in which the microwave power can be concentrated.

(3) A large magnet, to apply a d.c. field of about 3,000 gauss across the specimen.

(4) A detector to measure the power absorbed by the specimen at resonance.

(5) Suitable display and recording systems.

The arrangement of these different items to form the simplest type of room temperature electron resonance spectrometer is shown in

Figure 5. The source of the 3 cm wavelength radiation is provided by a standard radar klystron valve which feeds into a waveguide run. This leads to the cavity resonator which acts as the absorption cell. The cavity resonator has the property of concentrating a high level of microwave power in its interior, and the specimen to be examined is placed in the centre of the resonator in the region of maximum microwave magnetic field strength. The output waveguide from the cavity then leads to a crystal detector which produces a d.c. output voltage proportional to the level of microwave power

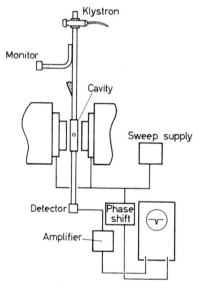

Figure 5. *Simple electron resonance spectrometer employing crystal video detection*

falling on it. The amplification and display systems can take various forms, but in the simplest of these the main d.c. magnetic field is modulated by a small a.c. component and the output from the detecting crystal is amplified by a wide-band audio amplifier and then fed to the Y plates of an oscilloscope. The X plates of this are fed in synchronism with the modulating field sweep, and in this way the resultant absorption line can be directly traced out on the oscilloscope screen. This system of display is known as ' crystal-video detection ', it is very easy to set up and simple to operate, but has a poor signal-to-noise ratio. The various methods of obtaining higher sensitivity, and more complete details of the microwave apparatus, are given in the next two chapters.

It can be seen that if the radical concentration is large enough to be detected by this simple system, a direct measure of the absorption line intensity and width can be obtained from the oscilloscope screen. Comparison of this line with that from a standard sample enables an absolute measurement to be made both of the integrated intensity, and thus the free radical concentration, and of the line width, and hence the relaxation processes affecting the electrons. It may be noted, in this connection, that the g-values of most free radicals remain very close to the free-spin value of 2·0023. This is because the electrons are delocalized and there is very little coupling between their spin motion and any orbital motion, and so, unlike most paramagnetic atoms, there is no extra contribution from the orbital momentum to the g-value. It is thus very difficult to employ any g-value variations as a method of identifying different radical species present in a mixture. On the other hand the same delocalization of the electrons, which renders identification by g-value variation difficult, often results in a very detailed and characteristic hyperfine splitting of the spectrum, and this can make radical identification very simple. The occurrence and analysis of the hyperfine splitting in free radical spectra is one of most powerful analytical tools of the electron resonance technique, and a few introductory remarks on the origin of this splitting will now be given.

1.7 HYPERFINE INTERACTION

If the orbit of the unpaired electron embraces an atom which has a nucleus with a magnetic moment and spin, then there will be an interaction between this nucleus and the electron, and the energy levels of the electron will be split by a small amount. If the particular case of interaction with one proton is considered it can be seen that the proton spin and moment will be lined up either parallel or anti-parallel to the applied field and electron spin. This is because the proton itself has a spin of $I=\frac{1}{2}$ and therefore only the two possible orientations with components along the field of $\pm\frac{1}{2}$ are allowed. For the general case of a nucleus with spin, I, there would be $(2I+1)$ possible orientations. The magnetic moment of the proton will produce a small additional field at the electron, and this will either add to, or subtract from, the effect of the external field, according as the proton is quantized with its spin parallel or anti-parallel to the field. The electron will therefore experience a magnetic field which is either slightly greater, or slightly less than

22

the applied field, and since the protons will be more or less equally distributed between the two orientations, each of the electronic levels will be split into two, as illustrated in *Figure 6*.

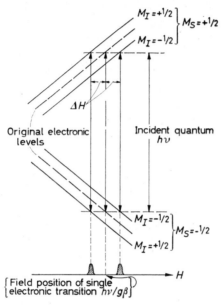

Figure 6.   *Splitting of electronic energy levels by interaction with one proton*

When radiation of the resonant frequency is applied the electron spins will again change orientation, and absorb the radiation but, in general, the nuclear spins will remain unchanged during the electronic transition. It therefore follows that transitions will only occur between levels which have the same quantized component of nuclear spin, and this selection rule can be expressed as $\Delta M_I = 0$. It can be seen from *Figure 6* that two transitions will therefore be obtained at magnetic fields given by

$$H_1 = \frac{h.\nu}{g.\beta} - \Delta H$$

$$H_2 = \frac{h.\nu}{g.\beta} + \Delta H$$

....(1.4)

For the general case of a nuclear spin, I, it can be seen that $(2I+1)$ component lines will be produced, with equal spacing between them. This nuclear interaction which splits the otherwise single electronic transition into several components is termed ' hyperfine interaction ' and the splitting between the lines is termed a ' hyperfine splitting '. It can be seen that this splitting is equal to $2.\Delta H$ and is in fact equal to twice the field produced at the electron by the proton magnetic moment. The magnitude of this splitting can be taken as a direct measure of the interaction between the proton and the unpaired electron, and will depend on how much of the electron orbital is concentrated near the proton. Thus, if the unpaired electron were in a $1s$ orbit of the proton a splitting of 500 gauss between the two component lines would be obtained, but if the electron were in a molecular orbital 90 per cent of which was spread over other atoms, the splitting produced by this proton would only be 50 gauss, and less still if it occupied higher orbits than the $1s$, when on the proton. A quantitative measurement of the hyperfine splitting will therefore give detailed information on the molecular orbital in which the unpaired electron is moving.

In most free radicals the unpaired electron is moving in a highly delocalized orbital which may embrace several nuclei possessing magnetic moments. Each of these will interact with the electron to produce a hyperfine splitting and the final spectrum may be rather complicated. There are two cases in which the spectrum can be very easily analysed, however, and, in practice, most free radical spectra fall into one or other of these two cases. The first is when the interaction with one nucleus, of spin I_1, is much larger than that with another nucleus, of spin I_2. The stronger interaction will then split the single electronic absorption line into $(2I_1+1)$ well separated lines, while the weaker interaction will further split each of these into $(2I_2+1)$ sub-components.

The second case of straightforward analysis is when the unpaired electron is equally coupled to n identical nuclei, of spin I. In this case an overlapping pattern of lines is obtained, the patterns all being shifted by a constant value of ΔH with respect to each other. The net result is an absorption pattern consisting of $(2n.I+1)$ lines with a maximum intensity at the centre and a symmetrical distribution on either side.

This type of pattern can be derived quite simply for the case of equal coupling with two or to three protons. Thus in *Figure 7(a)* the case of energy level splitting by hyperfine interaction with two protons is considered. Interaction with the first proton will split each of the electronic levels into two with $M_{I_1}=\pm\frac{1}{2}$ components, as shown. The further interaction with the other proton will split

24

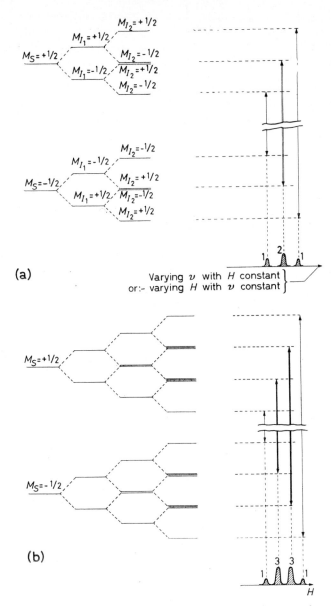

Figure 7. *Hyperfine splitting of the energy levels produced by two or three equally coupled protons.* (a) *Two protons;* (b) *Three protons*

25

each of these levels into another two with $M_{I_2} = \pm\frac{1}{2}$ components. Since the coupling and interaction with the two protons is assumed equal in this case, it follows that the $M_{I_2} = -\frac{1}{2}$ sub-component of the $M_{I_1} = +\frac{1}{2}$ component will be coincident with the $M_{I_2} = +\frac{1}{2}$ sub-component of the $M_{I_1} = -\frac{1}{2}$ component. As a net result each of the electronic levels is now split into three equally spaced sub-components. The selection rule is that both

$$\Delta M_{I_1} = 0 \text{ and } \Delta M_{I_2} = 0 \qquad \ldots (1.5)$$

and hence only three transitions are allowed, as shown. It may also be noted that, since the central transition is between levels which are doubly degenerate, the intensity of this transition will be twice that of the other two. Three equally spaced hyperfine components are therefore to be expected with an intensity distribution of $1 : 2 : 1$.

The case of equal interaction with three protons is shown in *Figure 7(b)*, and by following the same reasoning it can be seen that each electronic level is split into four equally spaced components, and a hyperfine pattern of four lines is thus obtained with an intensity distribution of $1 : 3 : 3 : 1$. It is also clear that the general case of n equivalent protons, with spin $\frac{1}{2}$, can be treated in this way, and the hyperfine pattern obtained will consist of $(n+1)$ equally spaced lines with an intensity distribution of

$$1 : n : n\,(n-1)/2 : \ldots n!/k!\,(n-k)! \ldots n : 1 \qquad \ldots (1.6)$$

where k varies from 1 to $(n-1)$.

Hyperfine patterns corresponding to an equal coupling of the unpaired electron with n protons are often to be found in the case of aromatic free radicals. The cases quoted above have in fact been directly recorded for such systems, thus the two-equivalent-proton case can be obtained from the spectrum of the 2.5.-dichloro-*p*-benzosemiquinone ion when a triplet with hyperfine separation between adjacent lines of 2 gauss is observed[32]. The unsubstituted *p*-benzosemiquinone ion has four equivalent protons, of course, and in this case[33] a pattern of five lines was obtained with intensity ratios of $1 : 4 : 6 : 4 : 1$.

It will be obvious from the above examples that the hyperfine structure of the resonance spectrum can be an exceedingly powerful tool in free radical identification and analysis. Since the lines are often quite narrow, overlapping spectra can usually be distinguished with ease, and the concentrations of different radicals, present in a system simultaneously, can thus be measured.

A more comprehensive treatment of the theory of the hyperfine interaction is given in the fourth chapter, and this first introductory section is only intended to serve as a background for the general theory and analysis presented later.

REFERENCES

[1] TAYLOR, N. W. *J. Amer. chem. Soc.* 48 (1926) 854
KENYON, J. and SUGDEN, S. *J. chem. Soc.* (1932) 170

[2] GOMBERG, M. Ber: 33 (1900) 3150; and *J. Amer. chem. Soc.* 22 (1900) 757

[3] PANETH, F. and HOFEDITZ, W. Ber: 62 (1929) 1335

[4] FREY, F. E. *Industr. Engng Chem. (Industr.)*, 26 (1934) 198

[5] SICKMAN, D. V. and ALLEN, A. O. *J. Amer. chem. Soc.* 56 (1934) 1251

[6] MULLER, E., MULLER-RODLOFF, I. and BUNGE, W. *Annalen*, 520 (1935) 235; 521 (1936) 89

[7] PEARSON, T. G., ROBINSON, P. L. and STODDART, E. M. *Proc. roy. Soc. A* 142 (1933) 275

[8] RICE, F. O. and JOHNSTON, W. R. *J. Amer. chem. Soc.* 56 (1934) 214

[9] TOLLEFSON, E. L. and LE ROY, D. J. *J. chem. Phys.* 16 (1948) 1057

[10] WATERS, W. A. *The Chemistry of Free Radicals.* Oxford University Press, 1946

[11] STEACIE, E. W. R. *Atomic and Free Radical Reactions.* Reinhold, New York, 1954

[12] ROY, M. F. and MARVEL, C. S. *J. Amer. chem. Soc.* 59 (1937) 2622

[13] WIGNER, E. *Z. phys. Chem.* B 23 (1933) 28

[14] SCHWAB, G. M. and AGALLIDIS, E. *Z. phys. Chem.* B 41 (1938) 59

[15] TURKEVITCH, J. and SELWOOD, P. W. *J. Amer. chem. Soc.* 63 (1941) 1077

[16] ELTENTON, G. C. *J. chem. Phys.* 10 (1942) 403; 15 (1947) 455

[17] HIPPLE, J. A. and STEVENSON, D. P. *Phys. Rev.* 63 (1943) 121

[18] ROBERTSON, A. J. B. *Mass Spectrometry.* Methuen, 1954

[19] LOSSING, F. P. *Ann. N.Y. Acad. Sci.* 67 (1957) 499

[20] LEWIS, G. N. and LIPKIN, D. *J. Amer. chem. Soc.* 64 (1942) 2801

[21] NORMAN, I. and PORTER, G. *Proc. roy. Soc. A* 230 (1955) 399

[22] INGRAM, D. J. E., HODGSON, W. G., PARKER, C. A. and REES, W. T. *Nature, Lond.* 176 (1955) 1227
GIBSON, J. F., INGRAM, D. J. E., SYMONS, M. C. R. and TOWNSEND, M. G. *Trans. Faraday Soc.* 53 (1957) 914

[23] NORRISH, R. G. W. and PORTER, G. *Nature, Lond.* 164 (1949) 658

[24] HERZBERG, G. and RAMSAY, D. A. *Disc. Faraday Soc.* 9 (1950) 80

[25] DAVIDSON, N., MARSHALL, R., LARSH, A. E. and CARRINGTON, T. *J. chem. Phys.* 19 (1951) 1311

[26] KOLTHOFF, I. M. and LINGANE, J. J. *Polarography.* Interscience, New York, 1952

[27] HORNIG, D. F. *Ann. N.Y. Acad. Sci.* 67 (1957) 463

[28] INGRAM, D. J. E. *Spectroscopy at Radio and Microwave Frequencies.* Butterworths, London, 1955, Chap. V

[29] DOUSMANIS, G. C., SANDERS, T. M. and TOWNES, C. H. *Phys. Rev.* 100 (1955) 1735

[30] SHARBAUGH, A. H. *Rev. sci. Instrum.* 21 (1950) 120

[31] GARSTENS, M. A., SINGER, L. S. and RYAN, A. H. *Phys. Rev.* 96 (1954) 53

[32] FRAENKEL, G. K. *Ann. N.Y. Acad. Sci.* 67 (1957) 546

[33] VENKATARAMAN, B. and FRAENKEL, G. K. *J. Amer. chem. Soc.* 77 (1955) 2707

[34] COSTAIN, C. C. *Canad. J. Phys.* 35 (1957) 241

[35] HURLE, I. R. and SUGDEN, T. M. *J. sci. Instrum.* 35 (1958) 450

EXPERIMENTAL TECHNIQUES

2.1 THEORY OF PARAMAGNETIC SUSCEPTIBILITY

BEFORE detailed consideration is given to the experimental techniques involved in electron spin resonance, the basic theory of paramagnetism is first outlined, together with a summary of the standard methods employed for measuring susceptibility.

2.1.1 Static or D.C. Susceptibility

The simplest case of a free radical, in which the unpaired electron is moving in a highly delocalized orbital, will be considered first. There will then be very little orbital contribution to the angular momentum and magnetic moment and, to a good approximation, only the spin need be considered. It has already been seen in *Figure 4* that, when a magnetic field is applied across such a specimen, the electrons will be divided into two groups, with an energy difference of $g\beta H$. It was also seen, from equation 1.3, that the ratio of the numbers occupying these two groups is given by the normal Boltzmann expression of $e^{-g\beta H/kT}$. In these expressions β is the Bohr magneton and equal to $eh/4\pi mc$.

Since the energies of normal microwave frequencies correspond to temperatures of about one degree absolute, the exponential can be approximated

to $\left(1-\dfrac{g\beta H}{kT}\right)$ for all experiments carried out at liquid hydrogen temperatures, $20° K$, or above.

If the total number of unpaired electrons per gramme in the specimen is N_0, it follows that there will be a net extra population of

the lowest level equal to $\left(\dfrac{N_0}{2}\cdot\dfrac{g\beta H}{kT}\right)$ electrons. In deriving this

expression it is assumed that $g\beta H/2kT\ll 1$. The full expression, with no approximation for the exponential, is $N_0\,(e^{g\beta H/kt}-1)/(e^{g\beta H/kT}+1)$,

which can also be written as $N_0\left(2\coth\dfrac{g\beta H}{kT}-\coth\dfrac{g\beta H}{2kT.}\right)$ In all

practical cases, however, the above approximation is justified and the susceptibility is given quite accurately by equation 2.1 below. Each of the excess electrons has a resolved magnetic moment equal to $\frac{1}{2}g\beta$ and there will therefore be a net magnetic

moment per gramme of the specimen equal to $\dfrac{1}{4}\dfrac{g^2\beta^2 N_0}{kT}\cdot H$. The susceptibility of the sample is defined as the magnetic moment divided by the field producing it, and the mass susceptibility per gramme is therefore given by

$$\chi_0 = \frac{1}{4\,kT}g^2\beta^2 N_0 \qquad \ldots (2.1)$$

If the unpaired electrons in the radicals were really ' free ' their g-value would be 2·0023 and the mass susceptibility would thus be

$$\chi_0 = \frac{0 \cdot 62 . 10^{-24}}{T} \cdot N_0 \qquad \ldots (2.2)$$

It can be seen from this that the number of free radicals present in a sample can therefore be determined very directly from the measured static mass susceptibility, and that

$$N_0 = 1 \cdot 60 . 10^{24} . \chi_0 . T \qquad \ldots (2.3)$$

where χ_0 is in c.g.s. electromagnetic units.

The same theory and equations will also apply if the electron has some orbital momentum and magnetic moment associated with it, but the g-value will now be changed from that of a free-spin.

The theory can also be simply extended for the case of two or more unpaired electrons existing at the same time in a molecule. This does not usually arise in free radical studies but often occurs in compounds of transition group elements. If the orbital contribution is ignored, and there are n unpaired electrons per molecule with their spins parallel, then the total spin quantum will be $S = \frac{n}{2}$, and there will be $(2S+1)$ quantized components of S in an applied magnetic field, acting along Oz. These $(2S+1)$ different levels correspond to the magnetic quantum number M_z varying from $+S$; $+(S-1)$; - - - to $-S$, and there will be an equal energy spacing between each of the $(2S+1)$ levels equal to $g\beta H$. The difference in population between the levels is again given by the Boltzmann distribution, so that the level characterized by the resolved component M_z will have a population of

$$\left(1 + \frac{M_z . g . \beta . H}{kT}\right) . \frac{N_0}{(2S+1)}$$

where N_0 is now the number of paramagnetic molecules per gramme.

The resolved magnetic moment of the aligned spins in this level is $M_z . g\beta$ and the net magnetic moment of all the spins in this level is therefore

$$(1 + M_z . g . \beta . H/kT)\frac{N_0}{2S+1} \cdot M_z . g . \beta \qquad \ldots (2.4)$$

29

The total magnetic moment of all the levels is therefore given by

$$\mu_{g.} = \sum_{M_z=-S}^{M_z=+S} \frac{N_0 \cdot g \cdot \beta \cdot H}{(2S+1)} \left[\frac{M_z}{H} + M_z^2 \cdot \frac{g\beta}{kT}\right]$$

$$= \frac{N_0 \cdot g^2 \cdot \beta^2 \cdot H}{kT} \cdot \frac{S(S+1)}{3} \qquad \dots (2.5)$$

since the first term sums to zero and for the second term

$$\sum_{-S}^{+S} M_z^2 = \frac{1}{3}S(S+1)\ (2S+1)$$

The static mass susceptibility is therefore now given by

$$\chi_0 = \frac{1}{3kT} \cdot N_0 \cdot g^2 \cdot \beta^2 \cdot S(S+1) \qquad \dots (2.6)$$

For the case of one unpaired electron per molecule, as in most free radicals, $S=\frac{1}{2}$ and equation 2.6 is identical to equation 2.1.

It may be noted, however, that equation 2.6 is the basis of all simple magnetochemical studies in which the number of unpaired electrons per molecule is ' counted ' by a static susceptibility determination. Since N_0 is known in such experiments the measured susceptibility effectively gives the value of $g^2S(S+1)$. This is the square of the magnetic moment of the aligned spins, and the value of $\sqrt{g^2S(S+1)}$ is termed the ' effective Bohr magneton number ', since it measures the magnetic moment in Bohr magnetons. If g is put equal to 2·0, this expression has the values of 1·7, 2·8, 3·9, 4·9 and 5·9 for 1, 2, 3, 4 or 5 unpaired electrons respectively. The valence state of the paramagnetic ion in a complex can there-fore be determined by taking the value of $\sqrt{\chi_0/(N_0 \cdot \beta^2/3kT)}$, as measured, and seeing with which of the above values it agrees best. It should be stressed, however, that all contributions due to electron orbital momentum are ignored in such a determination, and these may be quite considerable in some cases.

In free radical studies, however, the reverse procedure is usually adopted. Thus S is taken as $\frac{1}{2}$, with the effective Bohr magneton value as 1·7, or more accurately 1·732, and the value of N_0 is then determined from the measured χ_0, by use of equation 2.3. In this way the number of free radicals in a given specimen can be directly measured.

30

2.1.2 High Frequency and Complex Susceptibilities

In the preceding section the susceptibility of a paramagnetic specimen was derived assuming static conditions in a d.c. magnetic field. The case in which a high frequency field is also applied is now considered. Since the magnetization of a specimen will not follow a high frequency variation of the applied magnetic field instantaneously, but will have a phase lag associated with it, the general expression for the susceptibility can be written

$$\chi = \chi' - j \cdot \chi'' \qquad \ldots (2.7)$$

The real part of the susceptibility, χ', thus determines the in-phase magnetization, while the imaginary part χ'' determines the out-of-phase magnetization. If an oscillating magnetic field $H_1 \sin \omega t$ is applied across the specimen an in-phase magnetization of $\chi' \cdot H_1 \sin \omega t$, and an out-of-phase magnetization of $-\chi'' \cdot H_1 \cos \omega t$, will therefore be produced. The energy absorbed from the field is given by $H \cdot \dfrac{dM}{dt}$, and hence the total power absorbed is obtained by integrating over a cycle and multiplying by $\dfrac{\omega}{2\pi}$.

$$P_A = \frac{\omega}{2\pi} \int_0^{\frac{2\pi}{\omega}} H \cdot \frac{dM}{dt} \cdot dt = \frac{\omega}{2\pi} \cdot H_1^2 \int_0^{\frac{2\pi}{\omega}} \omega \cdot (\sin \omega t \cdot \chi' \cos \omega t + \chi'' \sin^2 \omega t) dt$$

$$= \tfrac{1}{2} \omega \cdot \chi'' \cdot H_1^2 \qquad \ldots (2.8)$$

It may be noticed that the first term integrates to zero and hence only the out-of-phase component, corresponding to χ'', absorbs power from the oscillating field.

This expression for the absorbed power is now equated to that obtained by considering the quanta absorbed during the electronic spin transitions. If the general case of n unpaired spins per molecule is taken, with a total spin quantum number S, then the difference in population between adjacent levels of M_z and $(M_z - 1)$ will be approximately

$$\frac{N_0}{(2S+1)} \left[1 + \frac{M_z \cdot \hbar \cdot \omega_0}{kT} - 1 - \frac{(M_z - 1) \cdot \hbar \cdot \omega_0}{kT} \right] = \frac{N_0 \cdot \hbar \cdot \omega_0}{(2S+1)kT} \ldots (2.9)$$

where ω_0 is the resonant angular frequency.

Each upward transition between these levels corresponds to an absorption of energy $\hbar \omega$ and if p_{m_z} is the transition probability,

31

the net absorption of power between these two levels will be

$$P_{M_z} = \frac{N_0 . \hbar^2 . \omega . \omega_0}{(2S+1) . kT} . p_{m_z} \qquad \ldots (2.10)$$

The value of p_{m_z} can be derived from standard radiation theory, and is given by

$$p_{m_z} = \frac{\pi}{4} \gamma^2 . H_1{}^2 (S+M_z)(S-M_z+1) . g(\omega-\omega_0) \quad \ldots (2.11)$$

where γ is the gyromagnetic ratio of the electron.

Here $g(\omega-\omega_0)$ is the 'line shape function' and represents the shape of the absorption line. It has a maximum at $\omega=\omega_0$ and is usually symmetrical on either side of ω_0. In most cases it approximates to a Gaussian or Lorentzian function depending on what interaction is the main source of broadening. Its value is normalized by making

$$\int_0^\infty g(\omega-\omega_0) . d\omega = 1 \qquad \ldots (2.12)$$

from which it follows that the maximum value of $g(\omega-\omega_0)$ is inversely proportional to the line width. The line width parameter can, in fact, be defined from this relation and the width of the line, as measured in frequency units, can be defined as the inverse of $\frac{1}{2}g(\nu_0)$. If the line shape factor is expressed and normalized as a function of the angular frequency ω, instead of ν, the expression for $\Delta\omega$ will be the inverse of $\pi . g(\omega=\omega_0)$, as illustrated in *Figure 35c*.

It can be seen that there will therefore only be an appreciable transition probability when the frequency of the applied magnetic field, ω, is close to the resonant frequency ω_0. This, in turn, implies that the specimen must also be in a strong d.c. magnetic field, H_0, such that

$$g . \beta . H_0 = \hbar . \omega_0 \qquad \ldots (2.13)$$

The expression for the power absorbed by transitions from the M_z-1 to the M_z level is therefore obtained by combining equations 2.10 and 2.11 to give

$$P_{M_z} = \frac{N_0 . \pi . \gamma^2 . H_1^2 . \hbar^2 . \omega . \omega_0}{4kT(2S+1)} . (S+M_z)(S-M_z+1) . g(\omega-\omega_0)$$
$$\ldots (2.14)$$

The total power absorbed by transitions between all the successive levels will therefore be given by

$$P_A = \frac{N_0 \cdot \gamma^2 \cdot H_1^2 \cdot \hbar^2 \cdot \omega \cdot \omega_0 \cdot \pi \cdot}{4kT(2S+1)} g\,(\omega - \omega_0) \cdot \sum_{M_z = -(S-1)}^{M_z = +S} (S^2 - M_z^2 + S + M_z).$$

$$= \frac{N_0 \cdot \gamma^2 \cdot H_1^2 \cdot \hbar^2 \cdot \omega \cdot \omega_0 \cdot \pi \cdot}{4kT(2S+1)} \cdot g(\omega - \omega_0) \cdot \frac{2}{3} S(S+1)\,(2S+1)$$

$$= \frac{\omega \cdot H_1^2}{2} \left[\frac{N_0 g^2 \cdot \beta^2 \cdot S(S+1)}{3kT} \right] \omega_0 \cdot \pi \cdot g(\omega - \omega_0) \qquad \ldots (2.15)$$

where γ has been replaced by $\dfrac{g\beta}{\hbar}$.

This expression is then equated to that of 2.8 to give

$$\chi'' = \left[\frac{N_0 \cdot g^2 \cdot \beta^2 \cdot S(S+1)}{3kT} \right] \omega_0 \cdot \pi \cdot g(\omega - \omega_0) \qquad \ldots (2.16)$$

It can be seen from equation 2.6 that the expression in the large brackets is the static mass susceptibility, χ_0, and hence

$$\chi'' = \chi_0 \cdot \omega_0 \cdot \pi \cdot g(\omega - \omega_0) \qquad \ldots (2.17)$$

If the measurements are made at the resonance frequency, ω_0, and the above definition of line width parameter $\Delta \omega = \dfrac{1}{\pi \cdot g(\omega_0)}$ is used then this equation can be written as

$$\chi''_{\omega_0} = \cdot \chi_0 \cdot \left(\frac{\omega_0}{\Delta \omega} \right) \qquad \ldots (2.17a)$$

It should be noted that this substitution is only justified if the 'spin temperature' is the same as that of the rest of the specimen, or 'lattice'. This will be so, to a good approximation, if the spin-lattice interaction is strong, but if this is weak, more power may be absorbed by the spins than they can dissipate to the lattice and thermal equilibrium between the two systems will no longer hold. This saturation of the energy levels is discussed in detail in Chapter 4, and it will be seen that, if this effect becomes appreciable, the power absorbed by the spin system will start to fall.

The important equation for use in free radical studies is really 2.15 as this relates the power absorbed in the cavity to the number of free radicals per gramme. Thus for $S=\frac{1}{2}$ it can be rewritten as

$$N_0=\left(\frac{8kT}{g^2 \cdot \beta^2}\right) \cdot \frac{\text{Power absorbed}}{\pi \cdot H_1^2 \omega \cdot \omega_0 \cdot g(\omega-\omega_0)} \qquad \ldots (2.18)$$

$$=1 \cdot 02 \cdot 10^{31} \cdot \frac{\text{Power absorbed}}{H_1^2 \cdot \omega \cdot \omega_0 \cdot g(\omega-\omega_0)} \cdot T \quad \ldots (2.19)$$

where the power absorbed is measured in watts and the microwave magnetic field in gauss.

The power absorbed is normally measured as a decrease in crystal current or the deflection of a pen-recorder. It is also to be noted that in all the above analysis so far it has been assumed that ω remains constant. In practice either the frequency or magnetic field is swept so that the whole absorption line is traced out. In this way the whole of the line shape $g(\omega-\omega_0)$ is integrated and is hence equal to unity, and if the observed 'power absorbed' is also equated to the integrated intensity under the absorption line, equation 2.19 can be rewritten as

$$N_0 \propto \text{Integrated intensity} \times \frac{T}{H_1^2 \cdot \omega_0{}^2} \quad \ldots (2.20)$$

provided that the line width is small compared with the resonant frequency.

The constant of proportionality now includes such factors as the gain of the amplifier system, and the units in which the intensity is measured. In practice its value is usually found by measuring the integrated intensity of a signal from a standard sample, rather than by substituting experimental amplifier gains and crystal conversion losses.

The variation of χ' and χ'' with frequency is considered in more detail later when discussing the various methods of detection and display for an electron resonance spectroscope. It can be seen from the above calculations, however, that either the static suscepti-bility χ_0, or the power absorbed due to the imaginary component χ'', can be used to measure free radical concentrations quantitatively. The experimental methods employed to measure these quantities are now described.

2.2 STATIC SUSCEPTIBILITY MEASUREMENTS

Nearly all methods of determining paramagnetic susceptibilities depend on a measurement of the force on the substance when it is placed in an inhomogeneous magnetic field. It can be shown from a simple vector analysis that the force on a magnetic dipole **m** has a component in the x direction of

$$F_x = m_x\left(\frac{\partial H_x}{\partial x}\right) + m_y\left(\frac{\partial H_y}{\partial x}\right) + m_z\left(\frac{\partial H_z}{\partial x}\right) \qquad \ldots . (2.21)$$

If this dipole is that produced in a paramagnetic substance of volume susceptibility χ and volume v, then its moment is equal to $\chi.v.H$ and

$$F_x = \chi.v.\left[H_x\frac{\partial H_x}{\partial x} + H_y\frac{\partial H_y}{\partial x} + H_z\frac{\partial H_z}{\partial x}\right]$$

$$= \frac{1}{2}.\chi.v.\left(\frac{\partial H^2}{\partial x}\right) \qquad \ldots . (2.22)$$

If the specimen has susceptibility χ_1 and is completely immersed in a medium of susceptibility χ_2 then the force will be given by

$$F_x = \frac{1}{2}(\chi_1 - \chi_2).v.\frac{\partial H^2}{\partial x} \qquad \ldots . (2.23)$$

This equation forms the basis of all of the static methods of determining paramagnetic susceptibility. The alternative methods vary in the different magnetic field configurations that are employed, and hence in the different values taken by $\frac{\partial H^2}{\partial x}$.

The most direct method of measurement is by the Gouy balance[1] in which the specimen takes the form of a long rod hanging vertically between the horizontal poles of the magnet. The specimen is made sufficiently long so that whilst one end is at the centre of the magnet gap, in a field H_1, the other end is well outside the gap, in a field $H_2 \approx 0$. The vertical force on any element of cross-section A is therefore

$$dF_x = \frac{1}{2}(\chi_1 - \chi_2).A.dx.\frac{\partial H^2}{\partial x} \qquad \ldots . (2.24)$$

and by integration over the length of the specimen, the total vertical force is given by

$$F_x = \frac{1}{2}A(\chi_1 - \chi_2)\int\frac{\partial H^2}{\partial x}.dx$$

$$= \frac{1}{2}A(\chi_1 - \chi_2).(H_1^2 - H_2^2) \qquad \ldots . (2.25)$$

The unknown susceptibility can therefore be determined from the values of magnetic field strength, and the value of F_x, as measured by a sensitive balance from which the specimen is suspended. The Gouy method is direct and can be applied to any solid or liquid specimen, provided large enough quantities are available. Liquids or powders can be placed in cylindrical glass tubes, which will not alter the resulting force, provided they are of uniform cross-section and extend equally on either side of the gap.

The Quincke method[2], which is a variant of the Gouy method, is often used for liquids. In this version the liquid is placed in a U tube which has one arm in the magnetic gap, and the other well removed from it. The force produced by the magnetic field is then balanced by the hydrostatic pressure between the different liquid levels in the two arms, and can be determined quite precisely by measuring the change in the height of the liquid with a travelling microscope. The use of this method has already been discussed in Chapter 1, and is illustrated in some detail in *Figure 2*.

The CURIE[3] and SUCKSMITH[4] balances are other variants which employ the same fundamental equation, but use specially shaped pole pieces in order to make $\frac{\partial H^2}{\partial x}$ as large as possible. The Curie method[3] is used for very small specimens, which are mounted on the end of a torsion arm in the region of maximum $\frac{\partial H^2}{\partial x}$, and the resultant force is measured by the torsion head. In the Sucksmith ring balance[4] the specimen is also suspended in the region of maximum field gradient and the force is measured by the distortion that it produces in a ring. This is pivoted vertically above the point of attachment to the specimen and has a mirror on either side of the pivot aligned to form an ' optical lever ' sensitive to any vertical elongation of the ring. A disadvantage of the Curie and Sucksmith methods is that it is difficult to ensure that different specimens are always replaced in the same position, and accurate measurement of the field gradients is not easy. They are therefore usually calibrated by a known compound, rather than used for absolute measurements. These difficulties may be partially overcome by employing more complicated systems such as the Curie–Cheneveau balance[5] in which the sample automatically places itself in the region of maximum attraction. These features are usually obtained however at the expense of sensitivity, and range of operating temperature.

A very sensitive balance for use with gases and vapours was developed by BITTER[6], in which a glass cylindrical vessel, divided radially into four equal chambers, is suspended from a torsion fibre between the poles of an electromagnet. Two of the chambers, diametrically opposed, are open to the gas, while the other two are evacuated. The two chambers containing the gas thus form an elongated paramagnetic specimen which will try and turn and align itself parallel to an applied magnetic field. The couple so produced, when the field is turned on, can be measured by the torsion of the fine quartz fibre used as suspension, and very small susceptibilities can be determined in this way. The method can also be used in a dynamic form since oscillations will occur if the specimen is displaced from its equilibrium position, and the period of these is a function of the magnetic susceptibility.

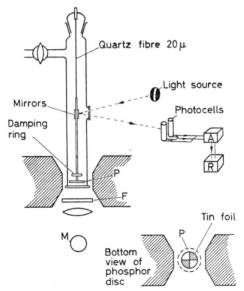

Figure 8. Detection of triplet states by d.c. susceptibility measurement. P, Phosphor disc; F, Filter; M, Mercury arc lamp; A, d.c. amplifier; R, Recorder (After Evans[7])

This form of balance has recently been adapted by EVANS[7] to study excited triplet state concentrations, and his apparatus is shown in *Figure 8* as it probably represents one of the most sensitive at present available for the determination of d.c. susceptibilities. The specimen holder now takes the form of a thin disc of the phosphor under investigation, about 1 cm in diameter and 0·5 mm thick, with alternate quadrants covered with tin foil. The back of the disc has a thin vertical rod leading up from its centre, carrying a small mirror and attached at its top end to the 20 μ quartz suspension fibre. The phosphor disc is illuminated from below by a mercury arc lamp, and the whole suspension is contained in an atmosphere of hydrogen gas. In this application the specimen is only paramagnetic whilst it is being illuminated and triplet excitation is produced. The pole faces of the magnet are shaped as shown to increase the couple produced by the field. The lens system and mirror form an optical lever which reflects the light on to two selenium photo-cells wired in series opposition, and connected to a push-pull amplifier and recording milliammeter. Rotation of the disc thus causes an unbalance d.c. signal which can be continuously recorded as a function of time and intensity of illumination. It was found that changes in susceptibility of 0·001 per cent could be detected by this method, which corresponds to less than 10^{-10} mole of triplet.

The sensitivity of the more direct methods of susceptibility measurement is very much lower than this rather specialized apparatus, however. On the other hand, the resonance measurements, in which χ'' is usually determined, enable high sensitivity to be obtained for a large variety of different sample types, and can also readily distinguish between the paramagnetic entity under investigation, and any ferromagnetic impurity or molecular diamagnetism. The wealth of extra information that can often be obtained from hyperfine structure also much enhances the value of employing resonance methods, instead of d.c. static methods, in most modern magnetochemical investigations.

2.3 THE BASIC ELECTRON RESONANCE SPECTROMETER

The general principles and basic experimental conditions for electron resonance have been briefly discussed in the first chapter. It was seen that, in theory, resonance could be obtained at any frequency, provided the applied magnetic field was adjusted to fulfil the resonance condition. Owing to the Boltzmann distribution in the energy levels greater sensitivity is obtained at the higher frequencies and fields, however, and a broad optimum frequency range of about 10,000 to 36,000 Mc/s occurs, above which suitable microwave generators and magnets are hard to find. Electron resonance spectrometers working in this range will therefore be considered first, although a considerable amount of work has been done on free radical spectra in the radiofrequency region and typical apparatus for this kind of work is discussed in the next chapter.

A simple microwave spectrometer has already been illustrated in *Figure 5*, and it was seen that the presence of the free radicals, or other paramagnetic entity, is shown by an absorption of power in the cavity when the magnetic field sweeps through the resonance value. In the transmission type spectrometer of *Figure 5*, this absorption of power appears as a small dip in an otherwise constant level, as is also illustrated by the output level variations in *Figure 9(a)*. In some methods of detection it is not very efficient to have a small change in a large constant level of power, and much better

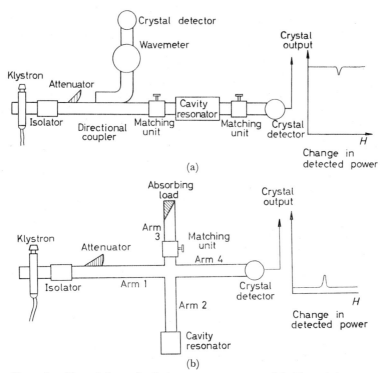

Figure 9. Transmission and reflection type spectrometers. (a) Transmission type, showing small proportional change in detected power; (b) Reflection type, showing large proportional change in detected power

results are obtained if the signal appears as a small positive power in the place of a zero mean level. This requires some kind of null-method or 'balanced-bridge' system, and a simple version of this is illustrated in *Figure 9(b)*.

This system has a four-arm element at the centre of the bridge, known as a 'magic-T' or 'hybrid-ring'. These are microwave components in which power fed into arm 1 is equally divided between arms 2 and 3, while none is fed into arm 4 unless a change in the matching of arms 2 or 3 occurs. The cavity resonator, containing the sample under investigation, is therefore placed at the end of arm 2 and matched with arm 3 so that when no absorption occurs no power is fed into arm 4. When the resonance condition is fulfilled, however, and power is absorbed in the cavity, an unbalance of the bridge will be produced, and hence power will then be fed into arm 4. If the detecting element is placed at the end of arm 4, the requirements for zero signal off resonance, and positive small signal on resonance, are therefore fulfilled. It will be seen later, in section 3.2, that this statement is really an over-simplification of the case since optimum sensitivity is not obtained when very small power falls on the detector. In practice, therefore, the bridge is never completely balanced, but sufficient unbalance is produced by the sliding screw matching unit to allow optimum power into the detecting arm. There is another reason for leaving a certain amount of unbalance, in that a mixture of both dispersion, χ', and absorption, χ'', will be fed to the detecting arm when resonance occurs, if the bridge is initially perfectly balanced. Either of these can be selected separately, however, if a certain amount of unbalance is purposely introduced.

These two simple types of microwave circuit, i.e. the transmission type and the balanced-bridge reflection type, form the basis of all electron resonance spectrometers. The former is more direct, and simpler to align, but the second can be used with more sensitive detecting systems, and can be employed to study both ' absorption ' and ' dispersion '. It will be seen that they have many microwave components in common, and before comparing their properties in a quantitative way, a brief description of the microwave components will be given.

2.4 MICROWAVE COMPONENTS

Nearly all electron resonance spectrometers employed for free radical investigations have used a wavelength of 3 cm, equivalent to a frequency of 9,000 Mc/s, which is a wavelength region commonly known as X-band. This is because free radical studies are usually concerned with detecting the minimum number of unpaired electrons *per unit volume*, in a specimen which has a relatively large size. The fact that much more of the specimen can be put in a 3 cm wavelength cavity, than in an 8 mm wavelength one, therefore considerably offsets the inherently higher sensitivity of the shorter wavelength. This may be expressed quantitatively in a simple way. It is shown in the next chapter, equation 3.25, that the minimum detectable susceptibility of a paramagnetic sample in a cavity is given by[8]

$$\chi''_{min.} = \frac{1}{Q_0 . \pi . \eta} \left(\frac{k T . \Delta v}{2 P_0} \right)^{\frac{1}{2}} \quad \quad \dots (2.26)$$

where η is the 'filling factor' and is equal to the ratio of the sample volume V_s, to the cavity volume, V_c, multiplied by a parameter which depends on the field distribution in the cavity and in the sample. If the cavity and specimen size are scaled with the wavelength in passing from 3 cm to 8 mm, this parameter will remain the same, and if the same electronic detecting system is used at both wavelengths equation 2.26 can be written as

$$\chi''_{min.} \propto \left(\frac{V_c}{V_s} \right) . \frac{1}{Q_0} \frac{1}{\sqrt{P_0}} \quad \quad \dots (2.27)$$

where Q_0 is the unloaded Q of the cavity and P_0 is the power incident on it. The number of unpaired spins will be proportional to $\chi_0 . V_s$ and hence from equation 2.17a

$$N_{0_{min.}} \propto \frac{V_c}{Q_0} . \frac{\Delta \omega}{\omega_0} . \frac{1}{\sqrt{P_0}} \quad \quad \dots (2.28)$$

If it is assumed that the same type of cavity is used at the two different wavelengths then the cavity volume decreases as ω_0^3 and the cavity Q decreases as $\omega_0^{\frac{1}{2}}$, so that, if the same power is available at both wavelengths, and the line width does not change

$$N_{0_{min.}} \propto \omega_0^{-\frac{7}{2}} \quad \quad \dots (2.29)$$

The sensitivity for detecting a small number of spins in a given small sample therefore rises very rapidly with frequency of measurement, if the above assumptions are true. In practice, however, the power available at the shorter wavelengths is smaller than that at 3 cm and the detecting systems are also inferior, but 8 mm wavelength spectrometers are nevertheless much more sensitive for studying specimens such as small single crystals.

If sensitivity for 'concentration of spins' is required, however, the quantity to be detected is the number of spins per unit volume,

i.e. N_0/V_s.

But

$$\frac{N_{0_{min.}}}{V_s} \propto \frac{V_c}{V_s} . \frac{1}{Q_0} . \frac{\Delta \omega}{\omega_0} . \frac{1}{\sqrt{P_0}} \quad \quad \dots (2.30)$$

The filling factor now remains constant and for the same incident power

$$\frac{N_{0_{\text{min.}}}}{V_s} \propto \omega_0^{-\frac{1}{2}} \qquad \ldots (2.31)$$

The sensitivity of the 3 cm wavelength spectrometer is now theoretically only twice as bad as that of the 8 mm wavelength version, and the lack of power and noisier detection systems of the latter in fact counterbalance this, so that the sensitivity of the two spectrometers is about the same. As mentioned in section 1.6, however, the practical considerations involving sample holders, ease of manipulation and the like, greatly favour the longer wavelength, and under these combined circumstances, the 3 cm band is invariably chosen for free radical studies.

Particular experiments at other wavelengths can be very useful, however, in differentiating between g-value variation and hyperfine splitting, and ancillary 1·25 cm or 8 mm wavelength spectrometers are very useful for this purpose. Their design is very similar to that of an X-band spectrometer, however, and only the 3 cm wavelength microwave components will be considered in detail here. If further description of the components used in the shorter-wavelength spectrometers is required it may be found in Chapters 2–4 of *Spectroscopy at Radio and Microwave Frequencies*[9].

(a) Klystrons

Klystrons working at about 3 cm wavelengths (9,000 Mc/s) can be divided into ' low voltage ' and ' high voltage ' types, and further subdivided by their output feed which is either ' coaxial ' or ' waveguide '. Each has its own particular advantage or disadvantage but for use as a source in an electron resonance spectrometer there is not very much to choose between them. If measurements on ' saturation ' and change of line width with input power are to be made then a high voltage klystron should be used as they have somewhat greater power output. These have the disadvantage, on the other hand, that it is much more difficult to apply automatic frequency control, necessary in superheterodyne reception, to a klystron with a reflector at $-2,000$ volts than to one with a reflector at -600 volts. In most high sensitivity spectrometers incorporating superheterodyne detection it is therefore usual to employ low voltage klystrons, when no difficult insulation problems occur in the control circuits.

Most low voltage 3 cm wavelength klystrons have been developed from the 723 A/B war-time X-band local oscillator. This works with about 350 volts between anode and cathode, and another 150 between cathode and reflector, with no separate grid control voltage. It has an electronic tuning range of about 50 Mc/s, a mechanical tuning range of about 10 per cent, and an output power of about 30 mW. These figures are generally typical of X-band low voltage klystrons, the E.E.V. K.300–335 and the Mullard type KS9–20A being examples of more recent English versions. A selection of typical klystrons is given in *Table 2.1*, with their operating data, but this is meant as only a representative rather than a complete list.

The high voltage X-band klystrons are mainly based on the CV 129 war-time version, and operate with an anode–reflector voltage of 2,000 volts, and, like all high voltage valves, have a separate grid potential to control the cathode current. They have an output power of about 50 mW, taking a current of about 10 mA instead of the 30 mA or so of the low voltage tubes. One version of the CV 129 has a particularly long tuning range, it is designated the CV 323, and is particularly useful if samples should change their dielectric constant markedly on cooling, or otherwise affect the resonant frequency of the cavity. It should be added that quite a large number of different X-band klystrons are now available on the American market, with high output powers and good stability. It should also be mentioned, however, that most of these are much more expensive than the European valves.

Table 2.1. Typical X-band Klystrons

Type	Manufacturer	Tuning Range (cm)	Output Power (mW)
Low Voltage 723 A/B	{ B.T.L. Raytheon	3·1 – 3·5	30
K.300–335	E.E.V.	3·1 – 3·2	25
KS9–20A	Mullard	3·1 – 3·5	35
V.A.6315	Varian	3·0 – 3·5	55
High Voltage CV 129	U.K. Services	3·1 – 3·3	50
CV 323	U.K. Services	3·0 – 3·5	50

The most important features of a klystron for use in an electron resonance spectroscope are (*i*) a *stable* output in both amplitude and frequency, (*ii*) a reasonable frequency range for tuning, and (*iii*) a low noise output in the detecting frequency band. In order to obtain as high a stability as possible, a well stabilized power pack must be used to supply the H.T. voltages, and the klystron filament should be run from batteries. Stabilized power packs suitable for low or high voltage klystrons can now be obtained from several different manufacturers (see Appendix), or detailed circuits may be found in such books as the M.I.T. series *Microwave Measurements*[10] and *Klystrons and Microwave Triodes*[11]. The klystron should be protected against draughts and sudden changes in ambient temperature, and the best way to accomplish this is to immerse it in an oil bath, with the output waveguide feed protruding from the top, and to water-cool or thermostat the oil-bath.

(*b*) *Attenuators*

These, as their name implies, are employed to reduce the microwave power and act by inserting a strip of resistive material into the region of maximum electric field strength. All microwave spectrometers employ a rectangular waveguide operating in the dominant H_{01} mode so that no other mode patterns should be propagated. This mode pattern in illustrated in *Figure 10*, where it can be seen that the lines of electric field are vertical and have a maximum along the central line of the waveguide, while the lines of magnetic field form closed loops in a plane at right angles to this. It is seen that a thin sheet of conducting material can be introduced into this region of maximum electric field either from a slot in the centre of the wide side of the guide or by a gradual movement over from a position parallel and close to the narrow wall. The former method was adopted in the earlier form of attenuators, while the latter is the more recent version and gives better accuracy in absolute calibration. Accurate measurement of attenuation is usually only required, however, if measurements on saturation are being made or an ' absolute ' determination of sensitivity is undertaken.

It may be mentioned here that there were initially two standard *X*-band waveguide sizes, namely the British (internal cross-section 1 in. $\times$ $\frac{1}{2}$ in.) and the American (internal cross-section 0·9 in. $\times$ 0·4 in.). The American size has now been accepted as the general standard and more microwave components are now available based on this waveguide than the other. It is therefore better to use the 0·9 in. $\times$ 0·4 in. inside cross-section for any new *X*-band spectrometer waveguide run. The wavelength of the radiation in the guide is not equal to the free-space wavelength, but is given by the relation

$$\frac{1}{\lambda_g^2} = \frac{1}{\lambda_{f.s.}^2} - \frac{1}{\lambda_c^2} \qquad \dots (2.32)$$

where λ_c is the cut-off wavelength and is equal to twice the width of the broad side of guide for an H_{01} mode. This fact must be borne in mind when designing any choke or matching elements, the size of which is determined by the wavelength.

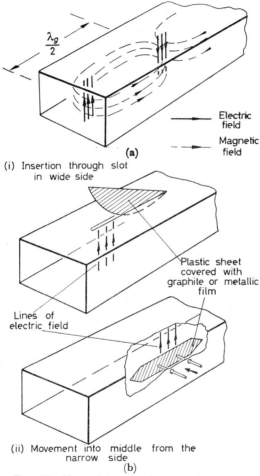

(i) Insertion through slot
in wide side

(a)

——— Electric field

- - - - Magnetic field

Plastic sheet covered with graphite or metallic film

Lines of electric field

(ii) Movement into middle from the narrow side

(b)

Figure 10. H_{01} mode in rectangular waveguide. (a) Electric and magnetic fields in a travelling wave; (b) Insertion of attenuating sheet

(c) *Isolators*

Isolators are a relatively new microwave circuit component and have the very useful non-reciprocal property of transmitting power with very little loss in one direction, but with very high loss in the other. This is accomplished by using Faraday rotation of the waves by a strip of ferrite material placed along the length of the guide and with a magnetic field applied across it to produce the right rotation for the wavelength concerned[12]. The planes of polarization of the input and output waveguide can then be adjusted so that waves travelling from input to output are rotated in such a way that their electric vector is always parallel to the narrow waveguide wall, while those travelling in the opposite direction have their plane of polarization rotated so that the electric vector emerges parallel to the wide side of the guide and the wave is therefore attenuated strongly.

These components can now be obtained from several manufacturers on both sides of the Atlantic and are usually produced as a ready-made unit incorporating a permanent magnet, and can be inserted directly and simply into the waveguide run. Care must be taken to ensure that they are inserted the correct way round so that they produce little attentuation of the power going from the klystron to the cavity, but cut out all the reflections from the waveguide system to the klystron. The insertion of such an element thus increases the amplitude and frequency stability of the klystron by a large factor, and they are essential for high sensitivity work requiring long time constants and a well stabilized microwave power source. They can often be inserted with advantage at other points in a microwave circuit, where power flow is only required to be unidirectional. As a particular instance, they can be placed in the main run just before additional power is added in a ' microwave bucking system ', and thus prevent unwanted reflections entering the cavity (see *Figure 21*).

(d) Directional Couplers

These components are designed to abstract a small proportion of power from the main waveguide run and feed it into an adjacent run so that the wave continues to travel in the same direction. They may take several forms but the simplest consists of two holes drilled in the narrow side of the waveguide, with their centres a quarter of a guide-wavelength apart. The magnetic field present along the narrow side of the main guide then couples power into the adjacent waveguide, and interference between the coupling at the two holes will take place, so that both add to form a wave travelling in the same direction as that in the main guide. The coupling can be increased by enlarging the size of the coupling holes, or reducing the thickness of the intervening wall. It can also be made more ' broad band ' by the use of staggered slots instead of holes.

46

Bethe single-hole couplers[13] can also be used. These have a single hole in the wide side of the guide and the directivity and amount of coupling is controlled by the angle between the two waveguides, and the size of the coupling hole. Two short stubs coupling the wide sides of the guides are also employed in another version, but the two-hole coupler outlined initially is the simplest, and usually the most satisfactory to use.

(e) Matching Screws

If a short stub, or screw, is partially inserted across the waveguide from the centre of the wide side of the guide it will reflect some of the incident waves and thus alter the standing wave pattern set up along the guide. In the low frequency equivalent circuit it acts as a series combination of capacity and inductance, and by adjusting its position and depth of penetration it can be made equivalent to any required reactance and can thus be used to compensate for any mismatch, or to balance a microwave bridge. It can be replaced by three stubs or screws fixed in position along the centre line of the guide, the depth of penetration of each of these being adjustable. It is probably easier to repeat balancing procedures by the use of a single movable screw, however, and this version is therefore to be preferred in microwave bridge circuits. Its main features of construction are outlined in *Figure 11*, and it is a matching unit used throughout electron resonance spectrometers.

(f) Cavity Resonators

The cavity resonator is the heart of the spectrometer, since it serves to concentrate the microwave power on to the sample placed at its centre, while it is at the same time held in the applied magnetic field. The ability of the cavity to concentrate the microwave power is measured quantitatively by its ' Q-factor ' which can be defined as

$$Q = \frac{\text{Energy stored in cavity}}{\text{Energy lost}} = \frac{\omega.\text{Energy stored}}{\text{Rate of energy loss}} \quad \dots(2.33)$$

If the energy lost is taken to be just that dissipated in the resistance of the cavity walls then the above ratio is called the ' unloaded Q ', while if the energy lost via the cavity coupling holes is also included, the ' loaded ' Q-factor is obtained. These Q values can be very high in the microwave region, of the order of 10,000 or more, although the presence of the sample in the cavity of a spectrometer lowers this value appreciably. It will be seen in the next section that the sensitivity of a spectrometer is directly proportional to the Q factor of its cavity.

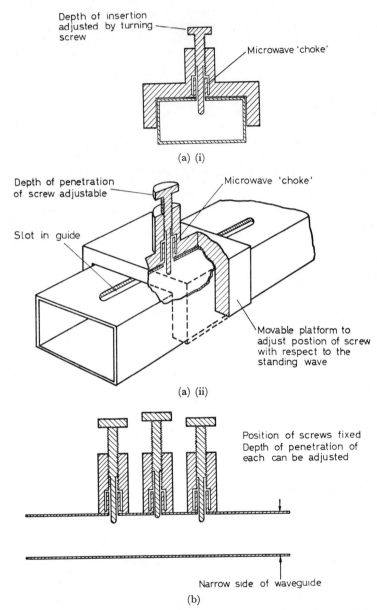

Depth of insertion
adjusted by turning
screw

Microwave 'choke'

(a) (i)

Depth of penetration
of screw adjustable

Microwave 'choke'

Slot in guide

Movable platform to
adjust postion of screw
with respect to the
standing wave

(a) (ii)

Position of screws fixed
Depth of penetration of
each can be adjusted

Narrow side of waveguide

(b)

Figure 11. Matching unit. (a) *Single movable screw* (i) *end view (cross-section);*
(ii) *Cut-away view;* (b) *Three fixed screws: side view (cross-section at centre)*

48

The high density of power within the cavity is produced by the high standing wave ratio, due to reflections at the ' shorted ' ends of the cavity. Thus the simplest type of cavity consists of a length of rectangular waveguide, an integral number of half-guide-wave-lengths long, and shorted by a metal plate at each end. This is illustrated in *Figure 12* where the resultant lines of magnetic and electric field are also plotted. For a stationary standing wave, such

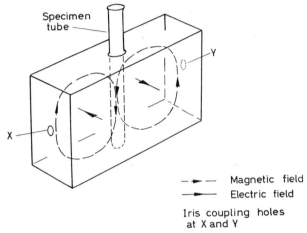

Specimen tube

Y

X

$- - \blacktriangleright - -$ Magnetic field

$\longrightarrow$ Electric field

Iris coupling holes
at X and Y

Figure 12. H_{012} mode in a rectangular cavity

as that in a cavity resonator, the maximum of the electric and magnetic fields are out of phase by $\pi/2$ in both space and time. The maximum electric field thus occurs $\frac{1}{4}$ and $\frac{3}{4}$ along such a cavity at the instant that the magnetic field is zero. In the case of a wave travelling along the guide the maxima of the two fields occur together and produce a power flow at right angles to both vectors.

The cavity of *Figure 12* is known as a rectangular H_{012} cavity. The H_{01} is the term given to the dominant mode propagated in the rectangular guide, as shown in *Figure 10*, and the subscript of 2 refers to the fact that two complete mode patterns, or half-wave-lengths, are included between the ' shorting ' walls. This desig-nation of cavity modes is quite general, thus an H_{mnp} mode is derived from an H_{mn} waveguide mode, in either rectangular or circular guide, and the subscript p denotes the number of half-wavelengths included in the cavity. In practice there will be small iris coupling holes in the centres of the end walls, as shown at X and Y, which couple the incident power in and can transmit a fraction of the stationary power out. The larger this coupling the more power is ' lost ' through the coupling holes and the lower the Q of the loaded cavity. It will be seen in the next chapter that the optimum coupling can be derived for any particular microwave system.

The requirement in an electron resonance absorption is that the sample should be placed in the region of maximum microwave magnetic field, since it is the magnetic component of the oscillating field that interacts with the electron moment. It is also important that the sample should be kept out of the oscillating microwave electric field as much as possible since this will produce a non-resonant absorption if the specimen has any dipolar character and drastically reduce the Q of the cavity. It is therefore evident that if a sample is to be inserted in the H_{012} cavity of *Figure 12* it should be placed vertically down the centre axis as shown by the dotted outline.

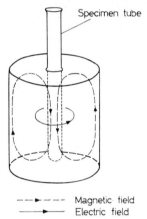

Specimen tube

$$\text{-- -- ► -- -- \quad Magnetic field}$$
$$\text{————► \quad Electric field}$$

Figure 13. H_{011} mode in cylindrical cavity

This H_{012} cavity is in fact used to a very large extent in free radical studies since field configuration is such that a glass or quartz tube can be inserted directly through the cavity as shown, without producing too much damping and loss of Q. This is ideal for liquid, powder or glassy samples, and most specimens containing free radicals are of this form. Another cavity with a field configuration very similar to that of the rectangular H_{012} is the cylindrical H_{011} as illustrated in *Figure 13*. (It may be noted that the zero subscript of a cylindrical waveguide mode indicates circular symmetry, and not zero variation of the field along the x direction as in the rectangular case.) This cavity again has its maximum microwave field along the central vertical axis and since this is now concentrated around a line, instead of a plane as in the rectangular case, a larger product of Q times filling factor (see equation 3.25) is obtained and hence a higher sensitivity. This cavity has the very considerable drawback that the cut-off wavelength for the cylindrical H_{01} mode is twice that of the dominant H_{11} mode and hence the cavity must have a diameter of at least 4 cm if it is to support a 3·2 cm wavelength H_{011} mode. It follows that the magnet gap must therefore be at least 4·2 cm, or considerably more if the cavity is to be cooled, and hence the homogeneity of the magnetic field may be considerably reduced. For this reason the cylindrical H_{011} cavity is not used very much for free radical studies when narrow absorption lines are expected.

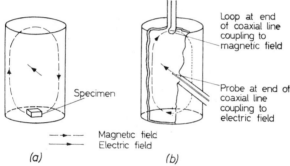

Figure 14. H_{111} mode in cylindrical cavity. (a) Field patterns; (b) Probe and loop coupling

The cylindrical waveguide mode with the smallest cut-off wavelength is the H_{11}, and an H_{111} cavity is illustrated in *Figure 14.* It will be seen that there is now only one closed loop of magnetic field in any given cross-section, and that the regions of maximum magnetic field are at the centre of the top or bottom faces. This cavity has the advantage that it is much smaller than the H_{011}, for any given wavelength, and can thus be readily immersed in a dewar and the whole inserted in a relatively narrow magnet gap. It is therefore used to a large extent in low temperature work, especially for single crystal studies, as these can be readily affixed centrally on the top or bottom. It has the disadvantage that no specimen tube can be inserted down its axis, however, since the electric field has a maximum across the centre. It is therefore difficult to insert much volume of material into this cavity, and it is not used to a large extent in free radical studies.

These three cavities are those most commonly employed in electron resonance spectrometers, and further details and modifications of them will be found in the sections dealing with actual spectrometers, and particular experiments, such as u.v. irradiation or heating *in situ*. The microwave power may be coupled into the cylindrical cavities by iris holes placed centrally in the vertical walls, or in the top face at regions of maximum microwave magnetic field. It has already been seen that cavities can be used either in a transmission, or reflection, system, as illustrated in *Figure 9*. It follows that two coupling holes and waveguide feeds are required for the former, whereas only one is needed for the latter. It is also possible to use coaxial feed instead of waveguide feed if there is a limitation of space, such as inside a dewar. The inner coaxial electrode can then either form a closed loop at the cavity input, when it couples to the magnetic field and should be placed in a region of maximum magnetic field; or it can form a projecting probe when it couples to the electric field and should be placed in a region where this has a maximum. These two types of coupling are illustrated in *Figure 14(b)* with respect to the H_{111} mode. Waveguide input is to be preferred, if it is possible, as it does not attenuate the microwaves so much as a coaxial line, and the amount of coupling is very much easier to adjust. If the cavity has to be immersed at the bottom of a dewar, however, coaxial line feed is often the only practical proposition, especially if a transmission system is being employed.

The use of low temperatures is often required in free radical studies since transient concentrations can often be ' frozen in ' and studied at leisure. Trapping processes, employing viscous hydrocarbon glasses at liquid nitrogen temperatures can also be very effectively used. The simplest method of cooling an H_{012} rectangular cavity is to use curved waveguide input and output sections, incorporating a region of ' heat break ', and immerse the cavity itself in a brass tank which can then be filled with liquid nitrogen. This arrangement is illustrated schematically in *Figure 15(a)*. The tank can be surrounded by a second metal container and the space between the two evacuated to act as a dewar. Alternatively the metal container can be packed around with one of the expanded plastic foam insulating materials, which have a very low thermal conductivity. For all temperatures down to $77° \text{K}$ this works very satisfactorily, and the system suffers from no vacuum leaks. The ' heat break ' consists of a length of guide about 10 cm long made of very thin brass or nickel-silver, supported by a moulded backing of paxolin or some other bad conductor, to give mechanical strength.

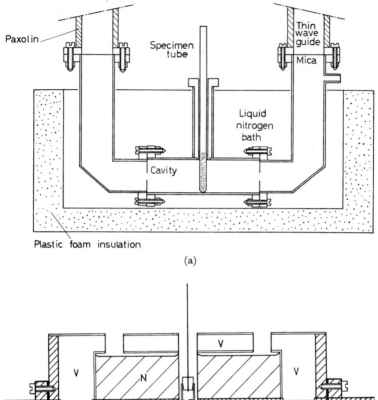

(a)

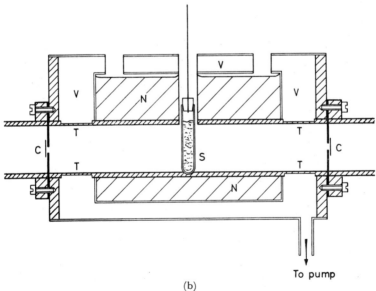

(b)

Figure 15. Low temperature cavities with direct waveguide feed. (a) *Demountable immersion type* ; (b) *Type with integral cavity and vacuum jacket. V, Vacuum. N, Liquid nitrogen. S, Specimen. C, Coupling holes with mica cover. T, Thin wall of waveguide*

53

Alternatively glass tubes coated with a thin silver film on the inside may be employed.

These features can be incorporated in a more elegant form by extending the waveguide of the resonator through the end walls of the liquid nitrogen container. The guide wall between the inner and outer containers, which form the vacuum jacket, is then milled down to about 0·01 mm thickness. This reduced wall thickness then serves as the heat break and the whole cavity can be operated in the H_{014} or H_{018} mode. The arrangement is outlined in *Figure 15(b)*. A more compact cavity and associated metal dewar is produced in this way, but it is likely to give vacuum troubles after short operation, and the more cumbersome but much more flexible system of *Figure 15(a)* is usually to be preferred.

More efficient heat breaks must be employed if temperatures below 77°K are to be used, however, and coaxial lines fabricated from nickel–silver tubing are normally employed as microwave feeds at liquid hydrogen temperatures and below. For reflection type cavities thin walled nickel–silver waveguide, or silver-coated glass waveguide can also be used, since one waveguide cross-section will normally fit into a dewar large enough to take the cylindrical cavity.

The cavities have been considered as of a fixed length so far, and thus tuned to a fixed frequency given, for the H_{mnp} mode, by

$$\nu = \frac{c}{\lambda_{f.s.}} = c \left[\left(\frac{p}{2d} \right)^2 + \frac{1}{\lambda_c^2} \right]^{\frac{1}{2}} \qquad \ldots (2.34)$$

where $\lambda_{f.s.}$ is the 'free-space' wavelength of the microwaves, d is the length of the cavity and λ_c is the 'cut-off' wavelength of the particular H_{mn} waveguide mode.

Cut-off wavelengths for the three different cavity modes considered here are given in *Table 2.2* and they are the longest wavelengths that the corresponding waveguide mode can propagate.

Table 2.2. Cut-off Wavelengths

Mode	Rectangular H_{01}	Cylindrical H_{11}	H_{01}
Cut-off Wavelength	2.a.	3·42.r	1·64.r

a is the width of the wider side; *r* is the radius of the cylindrical guide.

It is not very easy to alter the length of a rectangular cavity, as this means moving one of the rectangular shorting walls, but the cylindrical cavities can be easily tuned since the circular bottom face can be rotated and driven up and down in a screw thread. Details of this type of tuning, together with photographs of such cavities and tuning plungers may be found in *Spectroscopy at Radio and Microwave Frequencies*[14], but since these cavities are not employed to a large extent in free radical work they are not considered further here. It may be noted that the rectangular cavities can be tuned by inserting two small distrene stubs into the cavity at distances of a quarter of a guide wavelength from the shorting ends—i.e. into the region of maximum electric field. These should be spring-loaded and driven by a micrometer head for reproducible results. The cavities should be silver or gold plated if the largest Q values are to be obtained, and it is also best to have a set of shorting plates with different diameter coupling holes, so that the optimum value can be selected for various conditions.

(g) *Wavemeters*

Wavemeters are employed, as their name implies, to determine the wavelength, and hence the frequency, of the microwave radiation. They are, in effect, cylindrical cavity resonators with a movable bottom face so that the length of the cavity can be changed by known amounts. It can be seen from equation 2.34 that, if the mode in which the cavity is operating is known, then the wavelength of the radiation exciting it can be determined from the length of the cavity, d, that produces resonance. The cavity wavemeter can be used either as an absorption or transmission element. In the former case it is coupled into the side of the main waveguide, and when tuned to resonance it absorbs power from the main run and thus produces a sharp dip in the magnitude of the power flow down the guide. If used in transmission an output is taken from the cavity itself and power is only obtained in this arm when resonance occurs.

Any cavity mode can, in principle, be used for wavemeter operation, and a simple H_{11p} cylindrical cavity can be constructed with a central hole coupling its bottom face to the narrow side of the main waveguide run. This is the dominant mode and has the advantage that no other modes can be excited if the cavity diameter lies between $0 \cdot 58 \cdot \lambda$ and $0 \cdot 77 \cdot \lambda$. On the other hand this mode has not such a high Q value as some of the higher H_{onp} modes, and the accuracy of the wavemeter is directly proportional to the Q value of its cavity. For this reason the most accurate wavemeters are designed to operate with the circularly symmetric H_{onp} modes, when Q values of the order of 20,000 can be obtained. A diagram

55

of such a wavemeter is shown in *Figure 16* and it can be seen that the cavity is fed by two coupling holes on either side of the centre, and is used in transmission, the output guide being taken off at 45° to the main run. This type of input and output coupling prevents confusion with other modes such as the H_{11p} and H_{21p}. A typical diameter for such a wavemeter, operating at X band with a H_{01p} mode, is 5·0 cm, and the movable end wall of the cavity is driven by a precision micrometer. Greater accuracy is always obtained by taking two or three readings of the micrometer for successive resonances, averaging them, and then substituting this value of d into equation 2.34, with p equal to unity. In this way the 'end effects', inherent in a single measurement, are eliminated.

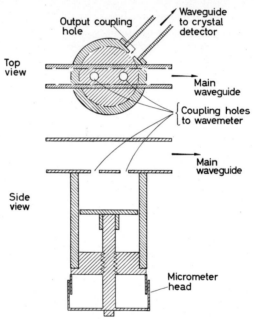

Figure 16. High-Q cavity wavemeter operating in H_{01p} mode

It may be noted that the measurement of wavelength, or frequency, is required in order to calculate the g-value of the absorption lines from

$$g = \frac{h\nu}{\beta.H} = \frac{21\cdot4184}{\lambda.H} \qquad \ldots.(2.35)$$

where λ, the free-space wavelength, is in cm and H is in kilogauss.

A cavity Q of 20,000 implies that the wavelength can only be measured to an accuracy of 1 in 10^4, and hence, if higher accuracies are required, the frequency must be measured directly by comparison with harmonics from a quartz crystal frequency standard. Details of such techniques are included in the description of complete spectrometers of section 3.3. Most free-radical spectra have 'g-values' very close to the free-spin value, however, and an accurate g-value determination can then be made by measuring the field difference between the centre of the given radical's absorption line and that of some standard radical, such as 1, 1-diphenyl-2-picryl hydrazyl, when both are observed at the same frequency, and, preferably, at the same time. The difference in g-value, Δg, between that of the particular specimen and the hydrazyl is then given by

$$\Delta g = \frac{\Delta H}{H_x} \cdot g_0 \qquad \qquad \ldots . (2.36)$$

where g_0 is the g-value of the hydrazyl, 2·0036, and H_x is the resonance magnetic field value for the given specimen. In this way the determination of new g-values is reduced entirely to a magnetic field measurement, and ΔH can usually be read directly off the recorder chart, this having been previously calibrated by proton resonance.

(h) ' Magic-T's and ' Hybrid Rings '

These have already been discussed briefly in section 2.3 when considering the microwave circuit of *Figure 9(b)*. One or the other of these is used in any ' balanced bridge ' or ' reflection cavity ' system, and both have the property that, when the second and third arms are correctly matched and terminated, then power fed into the first arm is transmitted equally to the second and third arms, and none to the fourth. They can thus be used to ' balance out ' the constant level of microwave power from the klystron, so that only the small change produced by resonance absorption is fed on to the detecting system in arm four. As mentioned earlier, the bridge system is never exactly balanced, and the optimum matching to be used in arm three is discussed in the next chapter. A diagram of a typical ' magic-T ' is shown in *Figure 17(a)*, and that of a hybrid ring in *17(b)*. The ' magic-T ' obtains its properties from the different polarizations which exist in its arms, whereas the hybrid ring depends on interference effects, with the path length between successive guides equal to a quarter of a guide wavelength.

The necessary condition for matching an arm of a microwave bridge is that it should produce no reflections. This may be fulfilled

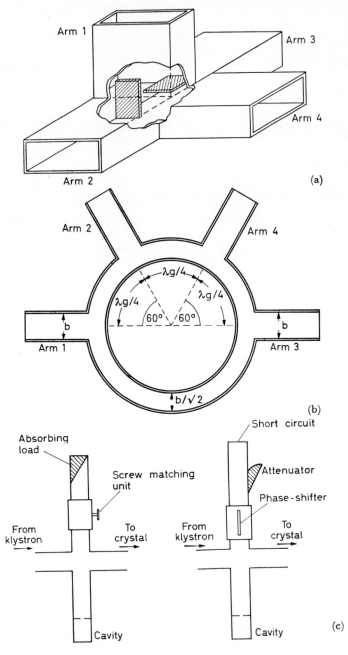

(i) Matching with adjustable
    screw and absorbing load

(ii) Matching with phase-shifter,
     attenuator and short circuit

Figure 17.   Microwave bridges.   (a) Magic-T;   (b) Hybrid-ring;
(c) Alternative matching for third arm

by inserting a perfectly-absorbing load at the end of the waveguide, and an approximation to such a load can be made from a tapered length of graphite-covered wood, of width equal to the narrow side of the guide. A small amount of residual reflection is usually produced, however, and it is wise to include a sliding-screw matching unit in the arm as well, to compensate for this. The waveguide arm leading to the cavity will be matched if the cavity is tuned to the microwave frequency and the coupling holes are of the right size to match the cavity resistive losses to the characteristic impedance of the guide (see section 3.1.1).

(i) Phase-shifters

Another microwave component that is often very useful is one that can alter phase but introduce no attenuation. Such phase-shifting devices can take two forms. The first consists of a length of guide which has an adjustable width, and allows λ_c and thus the guide wavelength to be altered. The second consists of a dielectric sheet, which can be gradually inserted into a central slot in the wide side of the guide, or moved across from the narrow side, as for an attenuator. These slow down the electric vector of the microwave field, and hence alter the velocity and wavelength of the microwaves in the guide. It can be seen that both of these devices operate on effectively the same principle and produce a shift in the phase of the standing wave pattern. The first type employs a long central slot in each of the two wide sides, and a screw clamp is used to force the narrow walls together. These ' slotted waveguide ' phase-shifters operate very well at 1·25 cm wavelengths and below, but the guide walls are too rigid at 3·2 cm wavelength for them to be very effective. Hence dielectric phase shifters are usually employed in X-band spectrometers, the depth of insertion being controlled by a micrometer drive.

It may be noted here that a very reproducible matching system for use in the balance arm of a microwave bridge can be made from a combination of a phase-shifter, variable attenuator and shorted waveguide end. This is sometimes to be preferred to the normal combination of a sliding screw matching element and absorbing load. Both forms of matching are illustrated schematically in *Figure 17(c)*.

(j) Crystal Holders

The crystal-holder, which terminates the detecting and monitor arms, must be designed so that the maximum amount of microwave energy is fed to the central electrode of the detecting crystal itself. A typical *X*-band crystal holder is shown in *Figure 18* and it is seen that a form of 'door knob' transition is placed across the wave-guide, which terminates in a small cylindrical tube into which the centre conductor of the crystal fits. This transition across the guide picks up the oscillating microwave electric field, and applies it as a change in potential to the silicon crystal contact. The position of a tuning plunger behind the transition is adjusted to produce a reflected wave which has the same phase as the incident wave at the transition and thus produces optimum coupling to the central conductor. It follows that the distance between the face of the shorting plunger and the crystal should be about a quarter of a guide-wavelength. A fixed-position three-screw matching element is also incorporated before the transition, so that the change of impedance at the latter can be compensated. It is seen that the rectified signal, at either the intermediate or modulation frequency, is fed out to a coaxial plug via a distrene support, which is designed to by-pass the microwave frequency components to earth.

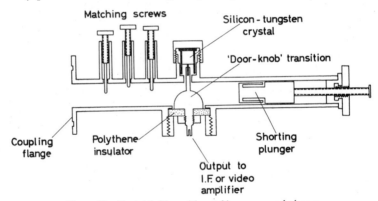

Figure 18. Crystal holder, with matching screws and plunger

Most modern *X*-band crystals are now of the coaxial shielded type as shown schematically in the crystal holder. These have internal impedances of the order of 400 ohms, and the variation of their noise and conversion loss is considered in detail in the next chapter. Manufacturers who supply *X*-band crystals are listed in the Appendix.

(k) Magnets

The magnet is the most expensive item in any electron resonance spectrometer, and the type of magnet required varies with the type of spectra to be studied. Requirements of field homogeneity are generally not so severe as those in 'High Resolution Nuclear Magnetic Resonance', but approach them in some studies, such as those of aromatic radicals in solution. Free radical investigations can, in fact, be divided into two groups from this point of view. If work is confined to radicals produced in an amorphous solid or glass phase, then absorption line widths greater than 2 gauss are usually obtained. It follows that a magnet with a homogeneity of about 0·1 gauss over the specimen may be used. Such magnets do not require extreme precision in construction, and the necessary homogeneity can be obtained with pole pieces of about 12 cm diameter, and with a U-shaped yoke carrying two energizing coils of standard construction. Such magnets can be purchased for the order of £300 and manufacturers of these are listed in the Appendix. The homogeneity of these magnets can always be increased by placing a raised rim around the outer edge of the pole faces, and detailed expressions for the dimensions of these may be found in papers by ROSE[15] and ANDREW and RUSHWORTH[16].

If free radicals in solution are to be studied, much better homogeneity may be required, however. The rapid tumbling motion of the liquid then averages out the dipolar broadening of the absorption lines, and very narrow lines may result. Hence inhomogeneities less than 1 milligauss may be required, and this can only be obtained from precision engineered magnets with large diameter pole faces. Unfortunately the cost of these magnets and associated power packs may be of the order of £5,000 or more, and it will be seen that the choice of magnet is therefore very largely controlled by financial considerations. In setting up an electron resonance spectrometer for general free radical studies it is best to obtain a magnet with as good a homogeneity as possible, since future work is more than likely to produce even narrower lines than those that have been observed so far. It should be pointed out, however, that there are still very wide fields of investigation, such as irradiation damage, trapping of radicals during polymerization, etc, which can be studied by those who cannot afford very high homogeneity magnets.

In electron resonance work on transition group compounds an electromagnet is always to be preferred to a permanent magnet, since wide g-value variations are often obtained which require changes of magnetic field by a factor of two or more. This requirement does not apply to free radical studies, however, where all the spectra will have g-values very close to the 2·0023 of free spins, and hence occur in a magnetic field of about 3,200 gauss for a spectrometer working at 3·2 cm wavelength. It therefore follows that a permanent magnet can be used in such work provided the value of

61

the magnetic field is accurately adjusted to 3,200 gauss during the initial magnetization. The advantage of a permanent magnet is that no power pack, with its associated stabilizing circuits, is required.

It will be seen later that the sensitivity of detection is inversely proportional to the bandwidth of the recording system. Hence for high sensitivity work the whole apparatus must be held very stable for several minutes while the absorption line is traced out. Any random small change of magnetic field during this time will distort the trace and may be mistaken for an extra hyperfine component. The main variation in magnetic field strength of a permanent magnet is due to temperature change and this is therefore usually quite gradual. Moreover, due to its large thermal capacity, the magnet can be relatively easily thermostated to within $0.01°C$[17]. The power pack supplying an electromagnet is liable to produce sudden changes of magnet current, however, due to varying input voltage and the like. It must therefore be very efficiently stabilized, to at least the same factor as the space homogeneity of the magnetic field, and the electronic circuits necessary for this can be very complex[18]. It is for this reason that the power packs for high homogeneity magnets often cost nearly as much as the magnets themselves. Electronic stabilization of magnet current can be most readily effected by valve circuits. Thin wire of relatively high resistance should therefore be used for the windings on the magnet coils so that high voltages and relatively low currents are employed. Stabilized power packs incorporating magnetic amplifiers are available for magnets with low resistance and high current windings, however (see Appendix), and electronic stabilizing circuits can also be devised[19].

It may be noted that sufficient stability can be obtained for magnets with homogeneities of about 0·1 gauss by feeding from a bank of batteries. Typical magnets of this type require about $3\frac{1}{2}$ amps at 50 volts, through two magnet coils of 3,500 turns 16 gauge wire, to produce 3,200 gauss. A stack of car batteries, charged overnight, and used during the day will therefore supply this, although it is wise to run the magnet for about half an hour before taking any spectra, to allow the battery current to settle.

(l) Magnetic-field Meter

The value of the magnetic field strength is determined by a proton-resonance meter which employs the nuclear resonance of the protons in the given magnetic field. Nuclear resonance is similar to electron resonance in principle, but the radiation couples to the nuclear magnetic moment instead of the electron moment. Since the value of the former is about two thousand times smaller than the latter the energy and frequency required to produce the transition

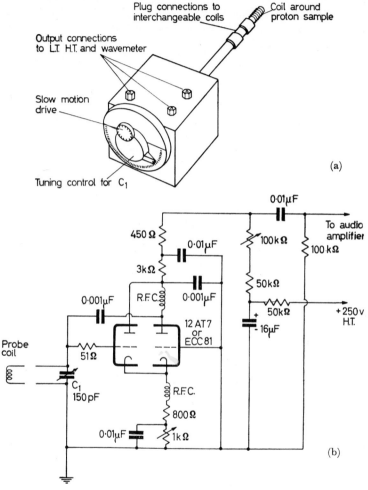

Figure 19. Proton resonance magnetic field meter. (a) *External appearance*; (b) *Circuit diagram*

of the nuclear moment is correspondingly reduced. Hence a resonance absorption of protons will occur in a field of 3,200 gauss at about 13·5 Mc/s, instead of the 9,000 Mc/s required for electron resonance. The gyromagnetic ratio of the proton has been determined with a high degree of accuracy[20] and the value of the magnetic field can thus be determined immediately from the measured proton resonance frequency. The numerical relation is

$$H = 2{\cdot}3487 . 10^{-4} . \nu . \text{gauss} \qquad \dots (2.37)$$

where ν is the frequency in c/s.

The specimen containing the protons, which may be a dilute aqueous solution of manganese sulphate or other paramagnetic salt, or simply glycerine, is placed in a small coil of about ten turns held at the end of a rigid coaxial line. The coil can thus be placed at the centre of the magnet gap while the other end of the coaxial line is connected to a variable condenser and suitable oscillator-detector circuit. The external shape and appearance of such a proton-resonance meter is shown in *Figure 19(a)* and the slow-motion dial and drive of the tuning condenser can be seen in the foreground. The circuit employed in such a meter is shown in *Figure 19(b)*, and consists of two cathode-coupled triodes which act as an oscillator-detector circuit. Amplitude modulation of the anode voltage of the second valve is produced when the protons absorb radiofrequency energy, and this can be fed out to an amplifier and display unit, as shown. The frequency of the radiofrequency oscillations can be measured by placing a small probe, or pick-up loop, inside the meter box, and bringing it out to a separate coaxial plug. This can then be connected to an accurate wavemeter and the frequency determined by heterodyne beating. If a large amplitude, low frequency magnetic field modulation is employed the proton absorption line can be displayed directly on the oscilloscope screen, and the wavemeter beat frequencies can also be seen.

It may be noted that the proton resonance signal can be used to achieve field stabilization[21] as well as field measurement. In this case a separate oscillating circuit is usually employed, and the output from the proton resonance head is fed to a phase-sensitive amplifier. This produces a d.c. output signal which is zero when the magnetic field is exactly that to give resonance, but swings positive or negative according as the field drifts too high or too low. Such an output signal can be used to activate a compensating current and to stabilize the field.

References

[1] Gouy, L. G. *C. R. Acad. Sci., Paris*, 109 (1889) 935
[2] Quincke, G. *Ann. Phys. u. Chemie.* 24 (1885) 347; 34 (1888) 401
[3] Curie, P. *Ann. Chim. Phys.* 5 (1895) 289
[4] Sucksmith, W. *Phil. Mag.* 8 (1929) 158

REFERENCES

[5] CHENEVEAU, C. *Phil. Mag.* 20 (1910) 357

[6] BITTER, F. *Phys. Rev.* 35 (1930) 1572

[7] EVANS, D. F. *Nature, Lond.,* 176 (1955) 777

[8] FEHER, G. *Bell. Syst. Tech. J.* 36 (1957) 449

[9] INGRAM, D. J. E. *Spectroscopy at Radio and Microwave Frequencies.* Butterworths, 1955, Chaps. II–IV

[10] MONTGOMERY, C. G. *Technique of Microwave Measurements.* McGraw-Hill, 1947

[11] HAMILTON, D. R., KNIPP, J. K. and KUPER, J. B. H. *Klystrons and Microwave, Triodes.* McGraw-Hill, 1948

[12] FOX, A. G., MILLER, S. E. and WEISS, M. T. *Bell Syst. Tech. J.* 34 (1955) 5

[13] BETHE, H. A. Radiation Laboratory Reports Nos. 194 and 199 (1943)

[14] INGRAM, D. J. E. *Spectroscopy at Radio and Microwave Frequencies.* Butterworths, 1955, p. 66

[15] ROSE, M. E. *Phys. Rev.* 53 (1938) 715

[16] ANDREW, E. R. and RUSHWORTH, F. A. *Proc. phys. Soc.* B.65 (1952) 801

[17] HUGHES, D. G. Ph.D. Thesis. Bangor, Univ. of Wales, 1957

[18] STRANDBERG, M. W. P., TINKHAM, M., SOLT, I. H. and DAVIS, C. F. *Rev. Sci. Instrum.* 27 (1956) 596

[19] ABRAHAM, R. J., OVENALL, D. W. and WHIFFEN, D. H. *J. Sci. Instrum.* 34 (1957) 269

[20] THOMAS, H. A., DRISCOLL, R. L. and HIPPLE, J. A. *Phys. Rev.* 75 (1949) 902

[21] PACKARD, M. E. *Rev. sci. Instrum.* 19 (1948) 435

THOMAS, H. A. *Electronics,* 25 (1952) 114

THE DESIGN AND CONSTRUCTION OF SPECTROMETERS

3.1 DESIGN OF MICROWAVE CIRCUITS*

IN this section the different factors which affect the sensitivity of an electron resonance spectrometer are considered at some length. The question of ultimate sensitivity is likely to play a large role in free radical studies, especially when electron resonance is applied to such systems as gaseous reactions, and employed to measure dynamic radical concentrations in polymerization, or radiation processes. A full consideration of the different factors which affect the ultimate sensitivity was therefore thought wise, especially as these principles will remain unchanged while the experimental results and their interpretations are likely to expand rapidly over the next few years.

The theoretical points which determine the design of the microwave circuit of an electron resonance spectrometer with optimum sensitivity are therefore now summarized. The mathematical derivation of the various optimum matching conditions will not be given at length, but the results will be quoted so that it can be seen which circuit elements must be adjusted in practice. A full treatment of this problem has been given by FEHER[1], and the derivation of the various conditions quoted in this section can be found in his paper.

The theoretical discussion can be divided into three parts:

(*i*) The derivation of the output signal (in terms of voltage or power) that will be produced by a given concentration of unpaired electrons.

(*ii*) The derivation of the maximum sensitivity, assuming noiseless detectors, and

(*iii*) the determination of the maximum sensitivity in the presence of detector and amplifier noise.

3.1.1 *The maximum output signal*

In this section the maximum change in voltage or power that can be produced by a given quantity of paramagnetic material is

---

* A fair knowledge of electronic theory is assumed in this section, but if the derivation of the conditions for maximum sensitivity are not required, the section may be omitted and the more practical considerations of 3.2 to 6 taken up directly.

DESIGN OF MICROWAVE CIRCUITS

calculated. The free radicals, or other paramagnetic entity, may be specified by a certain value of χ'', and a microwave output voltage of V_0, or power P_0, will be assumed from the klystron. The waveguide is assumed to have a characteristic impedance of R_0 and the cavity a resistive loss of r. The various Q factors of the cavity can then be defined as

Unloaded Q, i.e. the Q factor as determined by cavity wall resistance only

$$Q_0 = \frac{\omega L}{r} \qquad \ldots(3.1)$$

Coupled Q, i.e. the Q factor as determined by the coupling holes

$$Q_x = \frac{\omega L}{R_0 n^2} \qquad \ldots(3.2)$$

Loaded Q, i.e. the Q due to both cavity wall losses and coupling holes

$$Q_L = \frac{\omega L}{r + R_0 . n^2} \qquad \ldots(3.3)$$

In these definitions the low frequency equivalent circuit notation is employed, and L is the equivalent inductance which gives a resonant frequency ω and an unloaded Q of Q_0 for the given wall resistance of r. Similarly the coupling coefficient for the holes can be defined as Q_0/Q_x and is equal to $R_0 n^2/r$.

It may be noted that if the coupling coefficient is made equal to unity, the cavity will be 'matched' to the characteristic impedance of the guide. It will therefore produce no reflection, and this is the condition necessary for balancing the arm of a 'magic-T' or 'hybrid ring'. In practice the coupling holes are never of exactly the correct size, and a sliding screw matching element is usually incorporated in this arm as well, to compensate for any mismatch.

The extra loss produced by the resonance absorption of the paramagnetic sample can now be calculated as follows, from equation 2.8:

$$Q = \omega . \frac{\text{Energy stored}}{\text{Power Dissipated}} = \omega . \frac{\frac{1}{8\pi} \int_{V_c} H_1^2 . dV_c}{P_\omega + \frac{1}{2}\omega \int_{V_s} H_1^2 . \chi'' dV_s} \qquad \ldots(3.4)$$

where P_ω is the power dissipated in the cavity walls, V_c is the volume of the cavity and V_s that of the sample.

67

Since the absorption due to the sample is always small when high sensitivity is required, the denominator can be divided by P_ω and expanded to give

$$Q = Q_0 \left[1 - 4\pi \frac{\int_{V_s} H_1^2 . \chi'' dV_s}{\int_{V_c} H_1^2 . dV_c} . Q_0 \right] = Q_0 (1 - 4\pi . \chi'' \eta Q_0) \quad \ldots (3.5)$$

Hence the change in cavity Q factor due to the paramagnetic absorption is given by

$$\Delta Q = Q_0^2 \, 4\pi . \chi'' \eta \qquad \ldots (3.6)$$

where η is the filling factor and depends on the field distribution in the cavity and in the sample.

The next stage in the problem is therefore to express the change in voltage, ΔV, or in power, ΔP, in terms of ΔQ, and hence in terms of χ''.

The exact relation between these quantities will depend on what type of detector is being used, in particular whether its output is proportional to the power, or to the voltage, incident on it. If the detector output is proportional to the incident *power*, then the maximum power available from the source is calculated, followed by the change in reflected or transmitted power, when resonance absorption occurs (this may be represented as a change of δr in r,

provided $\left. \dfrac{\delta r}{r} = \dfrac{\Delta Q}{Q_0} \right)$.

The value of this change in power is then optimized by differentiation with respect to the coupling parameter, or n^2, and the ratio of the resultant change in power to the incident power can then be written in terms of ΔQ, and thus in terms of χ''. These steps may be summarized for the transmission and reflection type cavities as follows*:

Transmission	*Reflection*	
(a) *Power into cavity*		
$\dfrac{V_0^2 n_1^2 n_2^2 R_0}{(R_0 n_1^2 + r + R_0 n_2^2)^2}$	$\dfrac{V_0^2 n^2}{2} . \dfrac{r}{(R_0 n^2 + r)^2}$	$\ldots (3.7 \ \& \ 3.8)$

*These expressions follow from the method and calculations of G. Feher. *Bell. Syst. Tech. J.* 36 (1957) 449.

Transmission	Reflection

(b) *Change in power*

$$\frac{2V_0^2 n_1^2 n_2^2 R_0}{(R_0 n_1^2 + r + R_0 n_2^2)^3} \cdot \delta r \qquad \frac{V_0^2 n^2}{2} \cdot \frac{(R_0 n^2 - r)}{(R_0 n^2 + r)^3} \cdot \delta r \quad \ldots (3.9 \ \& \ 3.10)$$

(c) *Optimum coupling condition*

$$n_1^2 R_0 = n_2^2 R_0 = r \qquad \frac{R_0 n^2}{r} = 2 \pm \sqrt{3} \qquad \ldots (3.11 \ \& \ 3.12)$$

(d) *Expression for change in detected power*

$$\frac{\Delta P_T}{P_0} = \frac{8}{27} \cdot 4\pi \cdot \chi'' . \eta . Q_0 \qquad \frac{\Delta P_R}{P_0} = 0 \cdot 193 . 4\pi . \chi'' . \eta . Q_0$$
$$\ldots (3.13 \ \& \ 3.14)$$

It will be noted that the general expressions for power in the transmission type spectrometer contain two coupling coefficients, but that for the optimum value in change of transmitted power the input and output coupling should be identical.

The same kind of analysis can be made if the output of the detector is proportional to the incident voltage instead of the incident power. The change in the *voltage* reflected or transmitted is now calculated, and then optimized with respect to the coupling, and the steps can be summarized as before.

Transmission	Reflection

(a) *Voltage from cavity*

$$\frac{V_0 n_1 n_2 R_0}{R_0 n_1^2 + r + R_0 n_2^2} \qquad \frac{V_0}{\sqrt{2}} \left(1 - \frac{2 R_0 n^2}{R_0 n^2 + r} \right) \qquad \ldots (3.15 \ \& \ 3.16)$$

(b) *Change in voltage*

$$V_0 \left[\frac{n_2^2 R_0}{(r + 2 R_0 n_2^2)^2} \right] \cdot \delta r \qquad \sqrt{2} . V_0 . \frac{R_0 n^2}{(R_0 n^2 + r)^2} \cdot \delta r \ \ldots (3.17 \ \& \ 3.18)$$

(c) *Optimum coupling condition*

$$R_0 n_1^2 = R_0 n_2^2 = \frac{r}{2} \qquad \frac{R_0 n^2}{r} = 1 \qquad \ldots (3.19 \ \& \ 3.20)$$

(d) *Expression for change in detected voltage*

$$\frac{\Delta V_T}{V_0} = \frac{1}{8} \cdot 4\pi . \chi'' . \eta . Q_0 \qquad \frac{\Delta V_R}{V_0} = \frac{\sqrt{2}}{4} \cdot 4\pi . \chi'' . \eta . Q_0 \ \ldots (3.21 \ \& \ 3.22)$$

It will be seen that the numerical factors in equations 3.13, 3.14 and 3.21, 3.22 are all of the same order of magnitude, and hence, *from a consideration of the microwave circuit* there is no inherent advantage in using either transmission or reflection type cavities, or in using linear or square-law detectors, provided that the circuit is correctly optimized for the particular system that is employed.

The second step in the problem may now be taken—i.e. the derivation of the minimum detectable susceptibility, in the absence of detector noise.

3.1.2 *Minimum detectable susceptibility, with no detector noise*

The change in voltage or power at the end of the detecting arm is now assumed to act across an ideal matching resistor R_0 fixed at the end of the guide and possessing only thermal noise. The total root-mean-square noise voltage across this resistor will then be

$$V_{noise} = \sqrt{2} \cdot \sqrt{R_0 k T . \Delta \nu} \qquad \dots (3.23)$$

The factor of $\sqrt{2}$ arises from the fact that both the resistor and the noise from the waveguide system of internal impedance R_0 contribute to the total.

The minimum detectable susceptibility of the sample in the cavity will then be that which produces a value of ΔV which is just greater than V_{noise}. Hence, in the limit, taking the case of a reflection cavity and ΔV from equation 3.22

$$V_0 \cdot \sqrt{2} \cdot \pi \cdot \chi'' \eta \cdot Q_0 = \sqrt{2} \cdot \sqrt{R_0 k T . \Delta \nu} \qquad \dots (3.24)$$

$$\begin{aligned}
. \chi''_{min.} &- \frac{1}{Q_0 \cdot \eta \cdot \pi} \left(\frac{R_0 . k T . \Delta \nu}{V_0^2} \right)^{\frac{1}{2}} \\
&= \frac{1}{Q_0 \cdot \eta \cdot \pi} \left(\frac{k T . \Delta \nu}{2 . P_0} \right)^{\frac{1}{2}} \qquad \dots (3.25)
\end{aligned}$$

This equation therefore predicts the maximum sensitivity that can be achieved in any electron resonance spectrometer and assumes perfect, noiseless, methods of detection and display. It is interesting to insert figures into this expression and determine what is the absolute maximum sensitivity for a typical X-band spectrometer. Taking typical conditions, with $Q_0 = 5{,}000$, $P_0 = 10$ milliwatts, $\Delta \nu = 0 \cdot 1$ c/s, and for a free-radical of about 2 gauss line width, this gives a minimum detectable number of unpaired spins at room temperature of 10^{10}.

It can also be seen that equation 3.25 shows which parameters in the microwave circuit should be adjusted to obtain maximum sensitivity. Thus the sensitivity will rise linearly with an increase in Q_0 and with an increase in the filling factor. It will rise as a half-power factor of an increase in the incident power or of a decrease in the bandwidth of the detector. It should be stressed, however, that the noise due to the detecting system has not yet been considered and this may well affect the optimum value of these quantities, especially that of the microwave power.

3.1.3 *Maximum sensitivity, in the presence of detector and amplifier noise*

The final step in the analysis of the conditions necessary for maximum sensitivity is to include the noise of the detecting and amplifying systems. The ideal resistance, R_0, terminating the waveguide run, is now replaced by an actual microwave detector and associated amplifier which may employ either a crystal or bolometer as the sensitive element. The noise power at the output of this detector may then be written as [2,3]

$$P_{\text{noise}} = \left(\frac{N_m}{L} + F_A + t_D - 1\right) k \cdot T \cdot \Delta\nu. \qquad \ldots\ldots (3.26)$$

where N_m is the incident noise from the microwave run

L is the conversion loss of the crystal or bolometer

F_A is the noise figure of the amplifier

and t_D is the noise temperature of the crystal or bolometer.

(The 'noise temperature' of a circuit element is defined as the ratio of available noise power at the output of the element to that of an ideal resistor at room temperature.) The minimum detectable susceptibility will now be given by replacing the idealized $kT.\Delta\nu$ of equation 3.25 by the above expression for the noise power, and dividing the signal by the conversion loss, i.e.

$$\chi''_{\text{min.}} = \frac{1}{Q_0 \cdot \eta \cdot \pi} \cdot \left[\frac{(N_m/L + F_A + t_D - 1)L \cdot k \cdot T \cdot \Delta\nu}{2P_0}\right]^{\frac{1}{2}} \qquad \ldots\ldots (3.27)$$

This equation is quite general and will apply to all detection systems, it is the values of L, F_A and t_D which vary from system to system.

It therefore follows that maximum sensitivity will always be obtained by adjusting the product of the unloaded Q and the filling factor to be as high as possible, and then using a value of incident power which makes the expression in brackets of equation 3.27 a minimum, for the particular detecting system employed.

3.2 PRACTICAL METHODS OF DETECTION

There are only two types of microwave detector that are available in practice, at the moment, namely the silicon–tungsten crystal rectifier and the bolometer or barretter, which produces a change in resistance varying with input power. It is possible that in the relatively near future the low-noise travelling-wave tube detector or one form of solid state maser may supersede these but such are not yet available as reliable microwave components. The advantages and disadvantages of crystal as opposed to bolometer detection will depend on the particular conditions—i.e. microwave power, frequency of detection, and the variation of the parameters of equation 3.27 with these factors.

As a general statement it may be said that crystals have an excess noise which is inversely proportional to the frequency and increases with increasing incident power. Hence best signal-to-noise will be obtained with crystal detection if a modulating or intermediate frequency of high value is employed with relatively low input power to the crystal. Bolometer detection, on the other hand, has greater sensitivity when relatively high microwave powers are employed, as it has a high conversion loss at low powers, and high modulation frequencies are not required since there is no excess low-frequency flicker noise, as in a crystal. These different conditions will now be considered in more detail.

3.2.1 *Crystal Detection*

The best way to consider the sensitivity of crystal detection is to divide the actual minimum detectable susceptibility of equation 3.27 by that of equation 3.25, which assumes an ideal noiseless detector. The ratio, which measures in effect the loss of sensitivity by the detecting system is then given by

$$\frac{\chi''_{\text{min.} - \text{observed}}}{\chi''_{\text{min.} - \text{absolute}}} = [(N_m/L + F_A + t_D - 1)L]^{\frac{1}{2}} \quad \dots (3.28)$$

This expression will be termed the ' sensitivity loss factor ' and denoted by Δs.

It is therefore important to know how the conversion loss and noise temperature of the crystal change with different conditions. Crystal characteristics are roughly divided into two regions

(*i*) a square-law region, where the rectified current is proportional to the incident power, P_0; this holds at power levels below 10^{-5} watts,

(*ii*) a linear region, where the rectified current is proportional to $\sqrt{P_0}$; this holds for power levels greater than 10^{-4} watts.

The excess crystal noise is proportional to the square of the rectified current and inversely proportional to the frequency in both regions.

The conversion loss of the crystal is inversely proportional to the detected current in the square law region and constant in the linear region.

These characteristics may be summarized quantitatively by taking the proportionality constants as measured for a 3 cm wavelength 1N23 crystal by Feher[1] and writing

<table>
<tr><td align="center">Square Law Region</td><td align="center">Linear Region</td></tr>
<tr><td align="center">$< 10^{-5}$ watts</td><td align="center">$> 10^{-4}$ watts</td></tr>
</table>

Noise Power in watts

$$\left(\frac{5.10^{14}.P_f{}^2}{f}+1\right)kT.\varDelta f. \qquad \left(\frac{10^{11}.P_f.}{f}+1\right)kT.\varDelta f.$$

$$\ldots\ldots(3.29 \ \& \ 3.30)$$

Conversion Loss

$$L=\frac{1}{500.P_f} \qquad\qquad L\approx 3 \qquad \ldots\ldots(3.31 \ \& \ 3.22)$$

The power P_f, which is the power in the unmodulated carrier actually reaching the crystal, is measured in watts in the above expressions. These values may now be inserted into equation 3.28 to determine the quantitative decrease in sensitivity produced by the noise in the detector system. The noise figure for the amplifier is initially taken as unity, t_D is obtained from the expression for the noise power given above, and to a first approximation the noise from the microwave circuit N_m can be neglected. The only unspecified parameter is then f, and if a simple magnetic field modulation with external auxiliary coils around the magnet poles is envisaged the highest practical value for this is about 1,000 c/s. For such a simple direct form of detection the expression for loss of sensitivity of equation 3.28 then becomes

<table>
<tr><td align="center">Square Law Region</td><td align="center">Linear Region</td></tr>
</table>

$$\varDelta s = \left(\frac{1+5.10^{11}P_f{}^2}{500P_f}\right)^{\frac{1}{2}} \quad \varDelta s = [3(10^8.P_f+1)]^{\frac{1}{2}} \ \ldots\ldots(3.33 \ \& \ 3.34)$$

As an order of magnitude calculation the carrier power at the input to the crystal may be taken as about one tenth of the klystron microwave power, and the square-law region expression will then apply when $P_0 < 10^{-4}$ watts and the linear region expression when $P_0 > 10^{-3}$ watts. These two expressions can in fact be plotted graphically against microwave power, with the above assumption, to produce the curve of *Figure 20*. The two regions join at a minimum at around 5.10^{-5} watts of input power. It can be seen that the loss in sensitivity reaches a minimum of about 50, at the point where square-law operation changes to linear. The actual change is not so abrupt as the above analysis would imply but the physical reasons for this minimum are readily apparent— i.e. at low powers there is high conversion loss, whereas at high powers there is a large excess noise power in the crystal.

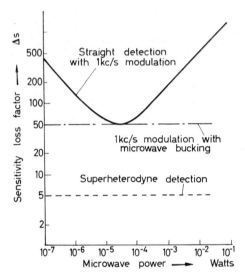

Figure 20. The effect of crystal noise on spectrometer sensitivity. The sensitivity loss factor is plotted against the microwave power (after Feher[1])

It would appear from the relatively sharp minimum of *Figure 20* that great care must be employed in designing the microwave run and coupling conditions so that the power incident on the crystal is of just the right value. This can be effected very easily in practice, however, by what is known as ' microwave bucking ', in which microwave power is artificially added to or subtracted from the detection arm. In the case of the reflection cavity circuit of *Figure 9(b)* this can be very simply effected by adjusting the sliding-screw matching element so that the bridge is taken further away from the ' balanced condition '. In the case of the transmission cavity circuit of *Figure 9(a)*, another waveguide run must be added to by-pass the cavity, as shown in *Figure 21*. This includes a phase-shifter and an attenuator so that both the amplitude and phase of the microwave power fed into the detecting arm at B can be altered. In this way the net level of power falling on the crystal can either be increased or decreased (phase difference of π) without altering the actual signal power, ΔP_T. It therefore follows that the optimum power condition can always be achieved by either adjusting the sliding screw of the reflection cavity, or the attenuator and phase shifter of the by-pass arm of the transmission cavity. This statement assumes, of course, that more than optimum power is available from the klystron, a fact which is true for all 3 cm wavelength work but may not be for the shorter wavelength spectrometers.

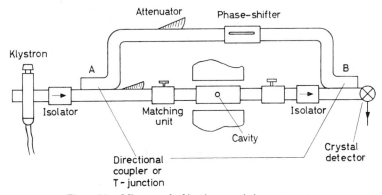

Figure 21. *Microwave bucking in transmission spectrometer*

The only way to decrease the sensitivity loss factor, Δs, below 50 is to increase f, the frequency of the detected output signal. It will be seen from equations 3.29 and 3.30 that the crystal noise power will fall as f rises until it becomes negligible compared with the other noise sources. This will happen at a frequency of about 10–20 Mc/s, and for this and higher frequencies the noise which limits the sensitivity will not be that of the crystal but of its accompanying intermediate-frequency amplifier (F_A of equation 3.26). Magnetic field modulation cannot be applied at frequencies as high as 30 Mc/s for two reasons

(i) it would be very difficult to produce a reasonably large sweep amplitude at the sample, and

(ii) modulation broadening of the absorption lines would be of the order of 10 gauss.

If the signal is to be detected at this frequency some form of super-heterodyne system must therefore be employed. This introduces a second microwave frequency into the detecting arm which beats with the original microwave power carrying the signal, to produce sum and difference frequencies. The 'local oscillator' frequency of the injected power is therefore adjusted to be 30 Mc/s higher or lower than the signal frequency, so that the crystal rectifies the signal as a 30 Mc/s beat frequency which is then fed into a standard narrow-band i.f. amplifier.

It can be seen that this method of superheterodyne detection has two inherent advantages

(*i*) the frequency of detection can be increased until the sensitivity is limited by the noise of the amplifier stages and not by the crystal, and

(*ii*) the power level falling on the crystal can be easily adjusted to optimum by varying the amplitude of the local oscillator input.

Referring back to equation 3.28, and equating t_D to unity for frequencies above 30 Mc/s the expression for the sensitivity loss now becomes

$$
\begin{array}{cc}
\textit{Square Law} & \textit{Linear} \\
< 10^{-5} \textit{ watts} & > 10^{-4} \textit{ watts} \\
\varDelta s = (N_m + F_A/500 P_f)^{\frac{1}{2}} & \varDelta s = (N_m + 3 . F_A)^{\frac{1}{2}} \quad \ldots (3.35 \ \& \ 3.36)
\end{array}
$$

Therefore, provided the local oscillator supplies more than 10^{-4} watts of power the sensitivity will be constant and at a maximum, with a minimum sensitivity loss of $(N_m + 3 . F_A)^{\frac{1}{2}}$. Substitution of typical figures for intermediate frequency amplifier noise and noise from the microwave circuit give a value of about 5 for this expression, and this value of $\varDelta s$ represents about the minimum sensitivity loss that can be secured with any practical microwave receiver to date. (Recent measurements on noise figures of 'masers' employing molecular amplification promise to improve this situation, but their practical operation has yet to be tested.)

It can be seen from the above analysis that if maximum sensitivity is required in an electron resonance spectrometer, superheterodyne detection should be employed. This has also been confirmed experimentally and the most sensitive spectrometers, detecting less than 10^{12} unpaired electrons, have been those using this system. A detailed description of such a spectrometer is therefore given in the next section so that those requiring the highest sensitivity can construct a spectrometer along these lines. It will become clear from that description, however, that such a system is somewhat complex and requires considerable adjustment, and it is therefore sometimes preferable to sacrifice the maximum sensitivity for a greater degree of flexibility, and employ the simpler system of high-frequency magnetic field modulation instead.

Although the frequency of magnetic field modulation cannot be increased to 30 Mc/s it can be increased considerably above the 1,000 c/s used in calculating the minimum sensitivity loss of 50 of equation 3.33. One of the limitations on the frequency used for modulation is purely practical—i.e. the difficulty of producing the required amplitude of sweep at the specimen, while the other is inherent in that any modulation of the absorption process will produce a line broadening approximately equal to the frequency of modulation[4]. The frequency of modulation should therefore never exceed about a tenth of the narrowest line width expected (as measured in frequency units).

In a large number of cases the narrowest width expected is determined by the magnetic field inhomogeneity, and the maximum permissible modulation frequency is therefore dependent on this. For example if the magnet available has an inhomogeneity of 0·2 gauss (600 kc/s) then the highest modulation frequency that can be used without producing any additional broadening is 100 kc/s. On the other hand, if free radicals trapped in the solid state are being observed, and it is known that the width of their absorption lines, due to dipolar interaction, will always be greater than 5 gauss (15 Mc/s) then modulation frequencies up to 1·5 Mc/s could be employed. The only case where modulation broadening is liable to be very noticeable is if free-radicals in solution are studied, with very homogeneous magnetic fields. The hyperfine components of radicals in solution are sometimes only 10 milligauss in width, and the frequency of magnetic field modulation must then be less than 5 kc/s. In general, it is probably fair to say that high frequency magnetic field modulation should only be used if either

    (*a*) the homogeneity of the magnet is not better than 0·2 gauss over the specimen, or

    (*b*) only free radicals in the solid state are to be studied, with widths greater than 1 gauss and no resolvable hyperfine structure.

Under these conditions, however, the use of high frequency magnetic field modulation can produce a spectrometer of high sensitivity which is very flexible to use and simple to retune or modify. It will be seen from equation 3.29 that for f equal to 100 kc/s the noise power from the crystal, at an input power of 10^{-5} watts, is already falling towards unity. It is therefore of the same order of magnitude as that from the other noise sources, and sensitivity loss factors as low as about 8 can be expected. Since this type of detection has been used in a large number of spectrometers constructed for free radical studies a detailed description of one is also given in the next section. It has the advantage over the superheterodyne system of being considerably easier to assemble and run, and less expensive to install.

At this point it may be well to consider the effect of the *amplitude* of the magnet field modulation on the sensitivity of detection. There are in essence, two ways in which magnetic field modulation can be employed. In the simplest system of ' crystal-video ' detection as described in section 1.6 the amplitude of the field modulation is large compared with the line width and the whole absorption is

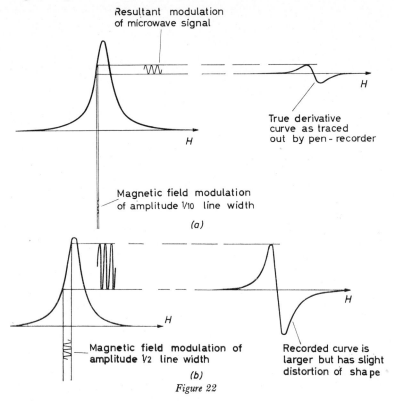

Resultant modulation
of microwave signal

True derivative
curve as traced
out by pen - recorder

Magnetic field modulation
of amplitude ⅒ line width

(a)

Magnetic field modulation of
amplitude ½ line width

Recorded curve is
larger but has slight
distortion of shape

(b)

Figure 22

swept through twice each cycle of modulation. The signal is thus in the form of a narrow pulse, repeated at twice the modulation frequency, and requires an amplifier of considerable bandwidth for faithful reproduction of the line shape. This system, although very simple and direct for large free-radical concentrations, is not easy to adapt for high sensitivity where narrow bandwidths are required to eliminate as much noise as possible.

In the second method, employed in all spectrometers using high frequency modulation, and in those incorporating phase-sensitive detection and pen-recording, the amplitude of the modulation is

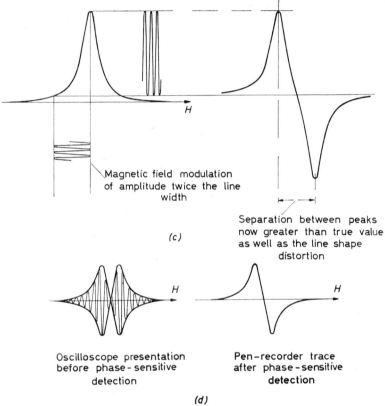

Magnetic field modulation of amplitude twice the line width

(c)

Separation between peaks now greater than true value as well as the line shape distortion

Oscilloscope presentation before phase-sensitive detection

Pen-recorder trace after phase-sensitive detection

(d)

Figure 22. *Small amplitude magnetic field modulation, as used in phase sensitive detection. The three curves show the signal that is obtained when a magnetic field modulation is swept slowly through the absorption line*

less than the line width. Its mean position is then moved slowly across the absorption line by a gradual change in d.c. current flowing round the magnet coils. The effect of this for three different amplitudes of sweep is shown in *Figure 22*, and it is seen that the small swing samples the slope of the absorption line and produces a signal level which is proportional to the derivative of the absorption.

The smaller the amplitude of sweep, the closer will the output follow the true derivative of the absorption, but, on the other hand, the smaller the amplitude of sweep, the smaller the actual output signal and thus the smaller the sensitivity. A compromise must therefore often be made between highest sensitivity (which is obtained with a sweep of the order of the line width), and display of true line shape (which requires a sweep less than a tenth of the line width). It may be noted, however, that whatever the value of the sweep, the centre of a symmetrical absorption peak will always be reproduced as the point at which the output crosses the axis (and hence the g-value can always be accurately determined). If the amplitude of the modulating field is known, the apparent increase in width that it produces can also be calculated. Hence, it is not too serious a matter if it is necessary to increase the sweep above a tenth of the line width, in order to obtain increased sensitivity. The equations derived in the previous section assume, in fact, that the full sensitivity, with the modulating amplitude equal to half the line width, is available for this method of detection.

3.2.2 *Bolometer Detection*

Bolometers have been used in several electron resonance spectrometers[5,6] instead of crystals, and quite high sensitivities have been obtained. They have the advantage that they possess no extra flicker noise at low frequencies and there is thus no need to take the frequency of detection up into the kc/s or Mc/s region. They have the drawback of a high conversion loss at low powers, however, and this, on the whole, renders them less acceptable than the crystal detectors.

The same type of analysis that has been followed for the crystal detector can be repeated for the bolometer, and a new expression drawn up for the sensitivity loss factor similar to that of equation 3.28. This has in fact been done by Feher[1], and details may be found in his paper. The general results are qualitatively similar to those for crystals, the main difference being that the minimum of the curve corresponding to *Figure 20* now occurs at 10^{-2} watts, instead of 5.10^{-5} watts. This difficulty can again be overcome by using a form of 'bucking' or 'balanced mixer' device to add extra microwave power into the detecting arm. The optimum conditions are more difficult to reproduce, however, since a relatively high level of power is required at the bolometer, and small changes in the matching stubs or phase changers can make a large difference in the signal. The earlier spectrometers[5,6] employing bolometers used a simple direct transmission system with no 'bucking' devices and relied on the fact that sufficient power was transmitted through the cavity to prevent too large a conversion loss in the bolometer. A detailed account of one of these spectroscopes is included in the next section since they form a very simple system with quite high sensitivity.

Practical tests[1] on various bolometers have shown that they usually possess more noise than that to be expected theoretically, however, and even when all the conditions are optimized the sensitivity loss factor does not approach very close to its minimum of 5. Hence superheterodyne detection, employing crystal rectification, is to be recommended whenever the highest sensitivity is required; and it is probably a fair statement to say that most of the laboratories in Great Britain which have tried bolometer detection have reverted to crystals sooner or later.

3.3 TYPICAL X-BAND SPECTROMETERS

The three spectrometers capable of high sensitivity, as discussed in the last section, are now described in detail. The high-frequency magnetic field modulation scheme is considered first since this is less complex than superheterodyne detection, and is probably used more than any other at the moment.

3.3.1 *High-frequency Modulation Spectrometer*

It should be stressed that this type of spectrometer should not be used if a magnet with a very high homogeneity is available and very narrow lines are expected. Considerations in the last section also show that the modulation frequency should not exceed 100 kc/s even when the field homogeneity is only 0·2 gauss. A spectrometer employing 100 kc/s modulation is therefore described, and this serves as a very useful high sensitivity instrument for all free radical studies in the solid, glass or polymer phases.

In low frequency magnetic field modulation the sweep is produced by a current flowing in auxiliary coils around the magnet pole pieces. It is not possible to use this technique at high frequencies, however, for two reasons

(i) very little sweep amplitude will reach the sample inside the conducting walls of the cavity (unless the latter is made of glass coated with a silver film less than the skin depth at 100 kc/s).

(ii) mechanical vibrations are set up in the walls of the cavity by an interaction of the induced eddy currents with the d.c. magnetic field. This produces a spurious modulation signal on the microwaves whenever the klystron is not quite

G

on tune with the cavity. Very large fluctuations in signal strength are therefore obtained for a small drift in klystron frequency. The effect can be overcome by the use of a glass cavity, as above, but this is often not very practical.

It is therefore necessary to produce the high frequency modulation by placing the modulating coil inside the cavity. A spectrometer incorporating this technique appears to have been first employed by BUCKMASTER and SCOVIL[7]. Their cavity was of the cylindrical H_{111} type and the cavity itself was made the modulating coil by inserting a small slot down its sides at the point where the microwave current flow was zero. This ' single loop ' modulating coil was fed with 465 kc/s, and they were able to detect less than

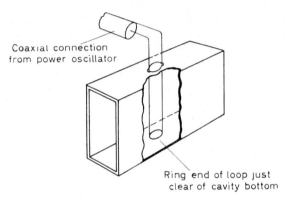

Coaxial connection from power oscillator

Ring end of loop just clear of cavity bottom

Figure 23. Modulating loop for 100 kc/s or 465 kc/s in an H_{012} cavity

10^{-11} moles of hydrazyl at room temperature. Later versions[8,9] of this type of apparatus in other laboratories have found it more convenient to use a separate single loop inside the cavity, and for the rectangular cavities used in most free-radical studies this is certainly to be preferred. A typical H_{012} cavity with modulating loop for 100 or 465 kc/s is shown in Figure 23. The loop is made of 20 s.w.g. enamelled cooper wire since relatively large currents will flow in it, and it also needs some mechanical rigidity. It is designed to fit closely on either side of a $\frac{1}{4}$ inch diameter specimen tube, and this is about the maximum size of sample holder that can be inserted into standard X-band rectangular guide. The loop is fed by coaxial line from the power output stage of a 100 kc/s oscillator via a matching transformer with only a few turns for its secondary. The circuit of a suitable oscillator is shown in Figure

24(*a*), and a volume control is included so that the sweep amplitude can be adjusted. A maximum value of about 5 gauss can be obtained with the circuit shown. Details of other types of high-frequency modulating loops may be found in a paper by LLEWELLYN[9]. These were confined to cylindrical H_{111} cavities, however, and used for single crystal studies.

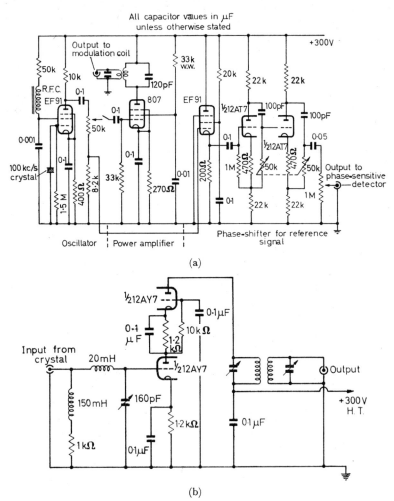

Figure 24. High frequency electronic circuits. (a) *100 kc/s oscillator for magnetic field modulation*; (b) *Low noise 100 kc/s preamplifier*

83

This cavity can be used in either the transmission or reflection systems of *Figure 9*. In either case the value of main magnetic field is slowly swept through resonance so that the microwave power is modulated at 100 kc/s. The modulation thus has an envelope proportional to the derivative of the absorption line, as illustrated in *Figure 22*. The output from the crystal detector is therefore fed to a narrow-band 100 kc/s amplifier, via a low noise pre-amplifier, typical circuit details of which are given in *Figure 24(b)*. A block diagram of a suitable detecting and display system, including phase-sensitive detection, with alternative oscilloscope display, is given in *Figure 25*. If the free radicals are present in sufficient concentration the output signal from the 100 kc/s amplifier can be displayed directly on the oscilloscope screen, when the derivative of the absorption line is seen modulating the 100 kc/s symmetrically.

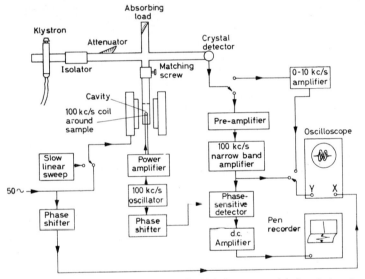

Figure 25. Detection and display system for high frequency modulation spectrometer

It should be noted that the output is not sensitive to the ' sign ' of the derivative until after phase-sensitive detection, and this accounts for the different appearance of the signal on the oscilloscope and recorder trace, as shown at the bottom of *Figure 22*.

In all studies requiring high sensitivity, phase-sensitive detection will need to be employed. This is a very convenient way of reducing the noise bandwidth so that only that of the final recording stage is effective. The time-constants of this final stage can be made of the order of seconds or minutes, so that the value of $\varDelta\nu$ (the inverse of the time constant) which is substituted into 3.25 or 3.27 is very much reduced from that required in oscilloscope display. The principle of phase-sensitive detection is to mix the 100 kc/s signal, modulated by the derivative of the absorption line, with one of constant amplitude but variable phase obtained directly from the oscillator producing the magnetic field modulation. The mixing can be effected either in a diode bridge circuit[10], or by feeding the signals to the different electrodes of one or two valves[11]. The operation may be viewed as a form of gating action[12] in which the signal amplitude is only allowed through the system by the peaks of the reference signal. A mathematical analysis shows that the output from the mixer contains the original modulation of the signal as a low frequency and this can then be fed into the narrow-band recording circuits, while the random noise beats of preceding stages are rejected since they are

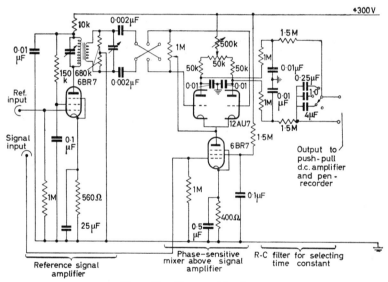

Figure 26. Phase-sensitive detector

not phase-coherent with the reference signal[13]. The circuit diagram of one typical phase-sensitive detector is given in *Figure 26*, and it is seen that the output is fed through a switched bank of *R-C* filters, which determine the time-constant of the final stage, to a d.c. amplifier and pen-recorder.

It will be seen that a switching control is included in *Figure 25* so that the whole detecting system can be changed to 'crystal-video' if required. This is usually employed when initially tuning the klystron and aligning the microwave circuit, and the high frequency modulation and detection system are only incorporated when the spectra are to be observed.

The slow sweep of the main magnetic field through the resonance line is usually produced by auxiliary coils (the same coils that are

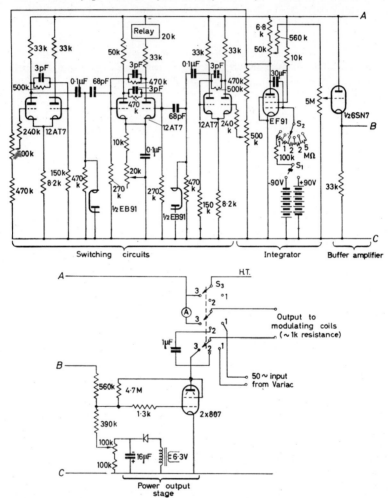

Figure 27. Electronic field sweep. This produces a sawtooth waveform of up to 5 minute period, with maximum current change of 120 mA

used to produce the 50 c/s modulating sweep required for crystal-video detection can be used). The d.c. current through these can be gradually changed, either by a mechanically-driven rheostat, or by an electronic circuit. The latter is preferable, since smoother variation in the field value is possible. A suitable circuit, based on a Miller integrator, is given in *Figure 27*. This produces a saw-tooth waveform with a period adjustable up to 5 minutes. The amplitude of sweep can also be controlled up to a maximum change of 120 mA, corresponding to a field sweep of 150 gauss.

Very slow sweep rates, with correspondingly long time-constants in the d.c. amplifier, will only produce greater sensitivity in the apparatus if the klystron frequency and current in the main magnet coils are constant throughout the time taken to record the spectra. Variations in klystron frequency are particularly trouble-some if a reflection type cavity is employed, since a signal proportional to χ' will be admixed with that proportional to χ'' when the cavity and klystron are not quite on tune. This distorts the spectrum and may lead to erroneous interpretation. It is therefore wise to use some form of automatic frequency control on the klystron. Pound stabilizer circuits[14], in which the klystron frequency is locked to that of a separate reference cavity, were tried in some initial spectrometers, but it is better to modify these and lock the klystron frequency on to the sample cavity itself[15]. This ensures that the klystron will follow any changes in resonant frequency produced by variation of temperature round the sample, or the like. Any dispersion signal, due to a change in χ', is thus automatically eliminated, and provided the cavity and klystron are thus held accurately on tune, slow sweep rates can be employed. A typical microwave circuit employing a modified Pound stabilizer is shown in *Figure 28*. An additional crystal is now placed at the end of arm 3 opposite the cavity, and this is fed with r.f. power from an oscillator of about 50 Mc/s frequency. Any microwave power travelling up arm 3 is thus modulated at 50 Mc/s and after reflection will return to the junction. Some of this will then return towards the klystron and be dissipated in the isolator, while the rest goes to the detecting crystal, is mixed with the unmodulated microwaves, and produces a 50 Mc/s output signal. The value of this signal is controlled by the matching conditions at the junction, and any change in klystron frequency with respect to the cavity will introduce an extra reactive component in this arm and thus alter the matching conditions. The sign of the reactive change will depend on the direction of frequency drift. This ' sign sensitiveness ' is preserved in the 50 Mc/s modulation so that after phase sensitive detection, the resultant signal can be used to raise or lower the klystron reflector voltage. The klystron frequency is thus readjusted to produce the same matching conditions as before, with the klystron accurately on tune with the cavity.

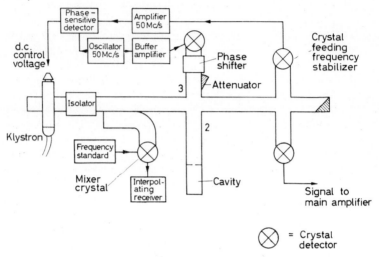

Figure 28.   Microwave frequency stabilizing circuit

Another, somewhat simpler, stabilizing system applies a small amplitude high-frequency sweep directly to the klystron reflector. If the cavity and klystron are exactly on tune no amplitude modulation of the microwaves is produced by the cavity, but any movement of the klystron frequency from the top of the Q curve will produce a sign-sensitive amplitude modulation in exactly the same way as illustrated in *Figure 22*. Such a signal can be filtered from the detecting crystal, and mixed with a reference signal from the oscillator in a phase-sensitive detector, the output of which controls the voltage on the klystron reflector. This system can produce modulation broadening, however, and is therefore not suitable when narrow lines are to be studied. It should be noted that any frequency stabilization scheme must have a response time at least as short as the inverse of the modulation frequency, since most of the detected noise components will arise here. It should also be stressed that any scheme which locks the klystron to the reference cavity, although very suitable for the detection of χ'', will be of no use for the detection of χ'. In most applications it is not necessary to detect changes in dispersion, and hence this is no drawback, but such measurements are sometimes required in saturation studies.

A frequency standard is also shown in *Figure 28*, feeding the crystal at the end of the monitor run. This consists of a thermostated quartz crystal operating at 5 or 10 Mc/s from which a series of multiplying chains derive a standard frequency at about 500 Mc/s. Some 25 Mc/s power, obtained from lower down the chain, is then added to modulate the 500 Mc/s signal, and the whole fed to the crystal mixer in the waveguide. The crystal acts as an efficient source of harmonics and also mixes one of these harmonics with the microwave power to produce a beat frequency in the high r.f. region. If the crystal is isolated from the guide the beat frequency can be fed directly to a calibrated radiofrequency receiver, and in this way the microwave frequency can be determined accurately to within a few kc/s. It is usually necessary to employ a cavity wavemeter as well, to determine which crystal harmonic is producing the beat frequency.

X-band high frequency modulation spectrometers incorporating stabilized klystrons and magnets are capable of detecting down to 2×10^{-11} moles of 1, 1-diphenyl-2-picryl hydrazyl (i.e. 10^{12} radicals with a line width of 1 gauss) when working at room temperature[9], with recorder time constants of 10 secs. This compares very favourably with the reported sensitivities of spectrometers employing superheterodyne or bolometer[5,6] detection.

3.3.2 *Spectrometers with Superheterodyne Detection*

The essential extra requirement for a system employing superheterodyne detection is another microwave source with a frequency about 50 Mc/s higher or lower than that of the main signal. This ' local oscillator ' frequency is injected into the detecting arm to beat with the signal frequency and produce the intermediate frequency of 50 Mc/s. It is also essential that the frequency difference between the two microwave sources should remain constant at the centre frequency of the i.f. amplifier, or otherwise its relatively narrow bandwidth may not pass the beat frequency. There are three ways in which these conditions can be met

(*i*) To employ very highly stabilized microwave sources so that there is no appreciable frequency drift of either the signal or the local oscillator klystron,

(*ii*) To employ two klystrons and incorporate ' automatic-frequency-control ' between them so that their frequency difference is held constant at the frequency of the i.f. amplifier; and (*iii*) To employ only one klystron, but abstract some of the power from the main waveguide run, modulate this by reflecting it from a crystal fed from a power oscillator at 50 Mc/s, and then inject the modulated microwaves back into the detector arm. The power in the side bands of the modulated input then act as the ' local oscillator ' frequency.

The latter technique is usually confined to work at shorter wavelengths than *X*-band, and either the first or second system employed for 3 cm wavelength spectrometers.   The first method, employing highly stabilized microwave sources, is, in principle, the

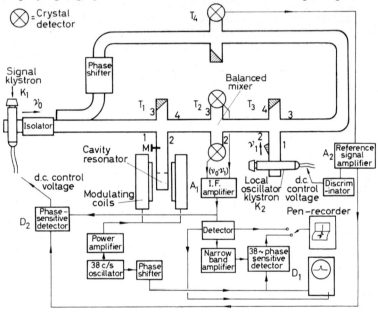

Figure 29.   Spectrometer with superheterodyne detection and frequency stabilization (After Hirshon and Fraenkel[16])

better, but such sources are very expensive to buy. Most workers are therefore faced with the construction of their own frequency-stabilizing system, and in this case the use of automatic-frequency-control between the two klystrons is usually to be preferred. It is also possible to lock the frequency of the signal klystron to the sample cavity at the same time, and thus ensure that only a variation of χ'' is detected as the signal.

A microwave circuit in which all these features are combined is shown in *Figure 29*, which is based on the spectrometer designed by

HIRSHON and FRAENKEL[16]. It will be seen that this circuit contains four ' magic-T ' or ' hybrid-ring ' bridge elements. The essential property of these is that, when correctly balanced, power fed into arm one (or four) will be equally divided into arms two and three, and none transmitted into arm four (or one). The klystron providing the main signal frequency, ν_0, is K_1 at the left, and this power is fed to the central magic-T or hybrid-ring, T_1, where it divides equally between arms two and three. The cavity containing the sample is placed at the end of arm two and the T can be balanced by the sliding screw matching element, M. Any absorption in the cavity will unbalance the bridge and a signal will be transmitted down arm four to the second magic-T, T_2. The ' local-oscillator ' klystron K_2, at the right of the figure, feeds the magic-T, T_3, so that its power is equally divided into arms two and three. The detecting magic-T, T_2, has crystals at the end of arms two and three, both of which receive the signal and local-oscillator power. The properties of the T are such, however, that whereas the signal power is of the same phase at each crystal, the local oscillator power has a difference of π in phase at the two crystals. Hence if the outputs of the two crystals are taken and mixed in a balanced transformer in the first stage of the intermediate frequency amplifier, the detected signals from the two sources will add, while the noise components from the local oscillator will cancel. This system of detection is known as a ' balanced mixer ' and is very effective in eliminating local oscillator noise.

The intermediate frequency $(\nu_0 - \nu_1)$ is amplified, detected, and the detected output mixed with a reference signal, from the oscillator supplying the magnetic field modulation, in the phase-sensitive detector, D_1. The field modulation is now at an audio frequency, to avoid modulation broadening. The output of the phase-sensitive detector, D_1, is then fed to a pen-recorder and the derivative of the spectrum traced out in the normal way. Alternatively, if large free radical concentrations are being studied a large amplitude field modulation may be used, and the output from the first detector fed straight to an oscilloscope. In this way the absorption line itself can be displayed directly on the screen.

The frequency-stabilization system is effected by mixing power from the local-oscillator klystron and the signal klystron in T_4. This is detected and fed to the i.f. amplifier, A_2, and the output of this will be independent of the resonance absorption and can thus be used as a ' reference signal ' source. This ' reference signal ' is mixed with an output taken from the signal i.f. amplifier, A_1, in the phase-sensitive detector. D_2, the output of which can be used to control the reflector potential of the signal klystron, K_1, and keep its frequency locked to that of the sample cavity. The frequency of the local oscillator is made to follow any changes in that of the signal klystron, so that their beat frequency remains constant, by also feeding the output of the reference amplifier to a ' discriminator ' which in turn controls the reflector voltage of the local oscillator klystron. A typical discriminator circuit is shown in *Figure 30* and is designed so

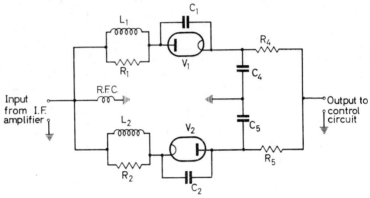

Figure 30. Discriminator circuit for automatic frequency control

that the combination of L_1 and C_1 will resonate at a higher frequency than the correct intermediate frequency, whereas the combination of L_2 and C_2 resonates at a lower value. D.c. potentials are thus built up across C_4 and C_5 of magnitude and sign corresponding to the deviation of the input frequency from the correct value. These changes of d.c. potential can then be fed to the klystron reflector to alter its tuning. The design of the phase-sensitive detectors and narrow band amplifiers is similar, in principle, to those discussed in the last section.

Although spectrometers such as this are rather complex, once they have been initially aligned, the adjustments required for new samples can be effected in a few minutes[16]. Reported sensitivities of this and other types of superheterodyne spectrometers[1,16] are of the order of 10^{-11} to 10^{-12} moles of 1, 1-diphenyl-2-picryl hydrazyl (10^{12} to 10^{11} free radicals of 1 gauss line width) for recording time-constants of 10 secs. This is only slightly greater than that of the best high-frequency modulation systems, but the superheterodyne spectrometers have the great advantage that they can be used to study very narrow lines, and produce no modulation broadening. It is also possible to build somewhat simpler microwave circuits containing only one balanced bridge, as in *Figure 9(b)*. The local oscillator power is then injected directly into the detecting arm and a balanced mixer feeds the i.f. amplifier. Automatic frequency control operates via a discriminator as above. The signal klystron can be locked to the cavity by a small amplitude voltage modulation applied to the klystron reflector, and any resultant amplitude

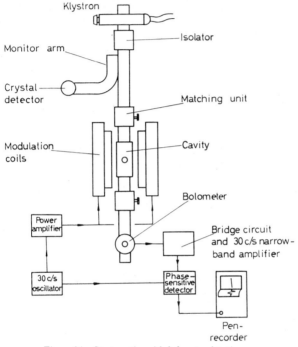

Figure 31. Spectrometer with bolometer detection

modulation produced by the cavity Q curve is amplified and phase-sensitive detected to control the d.c. reflector voltage, as outlined in the last section. This simpler version has the same order of sensitivity, but can suffer from modulation broadening, and if narrow free radical lines are to be studied with the highest sensitivity some circuit such as that in *Figure 29* is advisable.

3.3.3 *Spectrometers with Bolometer Detection*

Spectrometers employing bolometer detection have the simplest microwave circuit. The bolometer will work best with high microwave power falling on it (see section 3.2.2), and a direct transmission scheme as illustrated in *Figure 31* is normally used[5,6]. The bolometer is thus placed at the end of the waveguide run and incorporated in a d.c. bridge system, so that the change in voltage produced by its change in resistance can be readily fed to a narrow band amplifier centred on the frequency of the magnetic field modulation. Low frequency modulation, at about 30 c/s, is employed, to prevent modulation broadening, and also because the bolometer response falls off at higher frequencies, due to thermal inertia. Standard phase-sensitive detection and recording can therefore be used and a very simple and compact spectrometer of high sensitivity is thus obtained.

It should be noted, however, that bolometers vary considerably in their characteristics and even the best appear to have noise figures which are larger than expected[1]. A spectrometer with a good bolometer will have a sensitivity of about 10^{-11} mole of hydrazyl[6], at room temperature, however, and hence compares favourably with the other two forms of detection. The Sperry 821 barreter appears to have been the most successfully used in this type of spectrometer, but the names of other manufacturers are listed in the Appendix.

3.3.4   *Other Spectrometers*

Electron resonance spectrometers employing slightly different means of detection have been described in the literature. For example, ROSE-INNES[17] has used large amplitude frequency modulation of the klystron and detected the change in Q of the cavity, on absorption, by the reduction in height of a series of pulses. Other descriptions, apart from those already referred to, include the spectrometers of BERTHET[18]; RYTER, LACROIX and EXTERMANN[19] and UEBERSFELD[20].

Some work on free radical studies has also been carried out at radiofrequencies[21-23], mainly to investigate problems of line width and shape. There are two distinct disadvantages of working at radiofrequencies, however, i.e.

(*1*) lower inherent sensitivity

(*2*) the difficulty of interpreting any hyperfine structure when this has a splitting of the same order as the resonant field value.

Such work is therefore usually confined to the investigation of lines with small width and for these has the two compensating advantages of

(*1*) the fact that much larger volumes of *aqueous* solutions can be studied at these frequencies, and

(2) very high homogeneity magnetic fields can be produced at much less cost.

Radiofrequency spectrometers use the standard techniques of nuclear magnetic resonance, such as POUND and KNIGHT's[24] oscillator-detector circuit, or similar regenerative detectors[25], and usually incorporate phase-sensitive detection to obtain as high a sensitivity as possible. High homogeneity magnetic fields can be produced either by using large-radius Helmholtz coils, or by placing the whole apparatus in a solenoid. In either case allowance must be made for the direction and magnitude of the earth's field.

3.4   DOUBLE RESONANCE TECHNIQUES

Quite a new technique has recently appeared in electron resonance studies, known as the 'double resonance' or ENDOR method, and developed largely by FEHER[26,27]. This can be applied to any

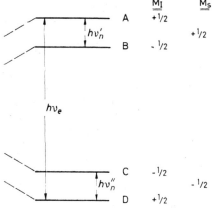

Figure 32. Precise energy levels of free radical electron interacting with one proton

spectrum in which a hyperfine splitting is to be expected, and consists of a simultaneous resonance experiment at a nuclear resonance frequency, at the same time as the normal electron resonance absorption. The principle of this may be illustrated in *Figure 32* where the energy levels of a free radical with one interacting proton are given in detail. Each of the electronic levels is split into two, and the energies of the four resulting levels will be:

Level	Energy	
A	$\frac{1}{2}g.\beta.H + \frac{1}{4}A - \frac{1}{2}g_N.\beta_N.H$	
B	$\frac{1}{2}g.\beta.H - \frac{1}{4}A + \frac{1}{2}g_N.\beta_N.H$	
C	$-\frac{1}{2}g.\beta.H + \frac{1}{4}A + \frac{1}{2}g_N.\beta_N.H$	.... (3.37)
D	$-\frac{1}{2}g.\beta.H - \frac{1}{4}A - \frac{1}{2}g_N.\beta_N.H$	

The last term in these expressions arises from the direct interaction of the applied magnetic field with the proton magnetic moment, and is very small compared with the second term. It follows, however, that the energy difference between levels A and B is not quite the same as that between C and D. The normal electron resonance condition is that $\Delta M_I = 0$ so that the electron resonance transition will be either between D and A or between C and B. In the former case the frequency will be given by

$$h\nu_e = g \cdot \beta \cdot H + \tfrac{1}{2}A \qquad \dots (3.38)$$

If relatively high power microwaves are fed to the sample at this frequency, this transition will tend to ' saturate ' so that the number of electrons in state A approaches that in state D. If, whilst this saturation is taking place, a radiofrequency, such that $h\nu_n'$ is equal to the splitting between A and B, $(\tfrac{1}{2}A - g_N \cdot \beta_N \cdot H)$, is also applied to the sample, stimulated transitions from A to B will be produced, and these two levels will become equally populated. As a result the electron resonance ' saturation ' is removed, and a strong line is obtained over the frequency region corresponding to the width of the nuclear resonance line. Similarly a frequency given by

$$h\nu_n'' = \tfrac{1}{2}A + g_N \cdot \beta_N \cdot H \qquad \dots (3.39)$$

will cause a ' desaturation ' of the electron resonance signal, due to transitions between D and C. It therefore follows that if the electron resonance frequency is held constant, and the radiofrequency is slowly swept about a value of $\nu_n \approx \frac{A}{2h}$, a large increase in the electron resonance signal will be obtained when $h \cdot \nu = \tfrac{1}{2}A \pm g_N \cdot \beta_N \cdot H$. These two values of ν_n can then be used to determine both A and g_N very accurately.

The more important application, so far as free-radical studies are concerned, is that the method can also be used to resolve hyperfine structure which has been lost in the electron resonance line width. This fact has been brought out very strikingly by the recent measurements of FEHER[28] on the structure of F centres formed in irradiated KCl. Normal electron resonance techniques had not been able to resolve any hyperfine structure, but by using the double resonance method Feher was able to resolve out the hyperfine splitting due to the Cl^{35}, Cl^{37} and even a trace of K^{41}. The spectrum he obtained is shown in *Figure 33* and it is to be noted that the base-line is now one of changing ν_n, and the electron resonance frequency and field are held constant all the time. The absorptions therefore represent a hyperfine *splitting* rather than a hyperfine line. The extra structure on the low frequency chlorine splittings is due to nuclear quadrupole interaction, and the detail available in this spectra, hitherto completely unresolved, indicates some of the possibilities of this new technique. A gain in resolution by a factor of 10^4 is in fact possible, since the limiting width is now that of the nuclear resonance instead of that of the electron resonance.

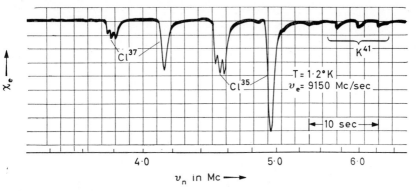

Figure 33. Double resonance spectrum of F centres in irradiated KCl. The sweep corresponds to a varying frequency of nuclear resonance (After Feher[28])

It may be noted that the principle behind this method is somewhat similar to that of the Overhauser effect[29], discovered a year or so before. In this the interaction between the nuclear and electron moments affects the spin-lattice relaxation times, and if the electron resonance is saturated the difference in population of the nuclei between their levels is greatly increased. The Overhauser effect can therefore be used to enhance the intensity of a nuclear magnetic resonance absorption by the simultaneous performance of an electron resonance[30,31], whereas the ENDOR technique can be used to improve the resolution of an electron resonance absorption by the simultaneous performance of a nuclear resonance[28].

3.5 MEASUREMENT OF INTENSITY AND LINE WIDTH

The determination of the absolute intensity of an electron resonance absorption can in principle be effected by the use of equations 3.13 and 3.14 and 2.17. Thus if the change in power occurring at the detecting crystal (or change of voltage as given by equations 3.21 and 3.22) is measured, as can be done relatively simply if the gain of the following amplifiers is accurately known, the complex susceptibility is given by (for a reflection cavity)

$$\chi'' = \frac{1.3}{\pi \cdot \eta \cdot Q_0 \cdot P_0} \cdot \Delta P \qquad \ldots \ldots (3.40)$$

Since η, Q_0 and P_0 can also be measured the absolute value of χ'' can be calculated in this way, and hence also the number of free radicals, from equations 2.17a and 2.3.

Such an absolute determination of free radical concentrations is seldom made in practice, however, and a comparison method, employing a standard sample, is adopted. For the best result the standard comparison sample should have a line width and intensity of the same order as those of spectra to be measured. It is then possible to trace out both spectra with identical conditions and gain-settings of the apparatus. The stable 1, 1-diphenyl-2-picryl hydrazyl has often been used as such a standard, since each molecule has one unpaired electron and it therefore possesses $1 \cdot 53 \cdot 10^{21}$ unpaired electrons per gramme (molecular weight $=394$). It should be noted, however, that if this is recrystallized from chloroform or similar solvents it will take up chloroform molecules of crystallization, and the molecular weight must be adjusted for these.

Measurement of small free radical concentrations by comparison with a hydrazyl sample becomes very difficult, however, since 1 microgramme of hydrazyl possesses 10^{15} unpaired electrons. It is therefore best to employ a subsidiary 'standard sample' for measuring radical concentrations of 10^{15} per gramme or below. Those working in the author's laboratory have found carbon samples, formed at about $500°$ C, very suitable for this purpose. Such carbons contain a large number of trapped free radicals[32,33], and can be readily ground and then diluted with an inert powder, such as a metal oxide, to produce samples of any required radical concentration. Care should be taken to test the inert powder for any paramagnetic impurities, however, before making up the standard samples. If they are kept in sealed tubes these samples remain very stable, and a whole series of concentrations ranging from 10^{16} electrons per gramme down to 10^{12} electrons per gramme can be made in this way by successive dilution. They are cross-checked and the higher concentrations checked against hydrazyl samples at regular intervals. The advantage of using carbon samples is that these can be finer ground and more evenly distributed in the inert filler, than hydrazyl itself, and the samples have a line width of about 8 gauss which is very similar to that of a large number of free radicals observed in the solid state. Such comparison samples, filled into standard quarter-inch diameter specimen tubes, can be inserted before and after the spectra of unknown intensity have been recorded, often without retuning or rematching the waveguide run, and the recorder traces can then be compared directly.

It should be noted that if phase-sensitive detection is employed, the curve must be integrated first to obtain the absorption curve, and this then integrated again to obtain the overall absorption. Electronic circuits can be built to perform the integration at the same time as the derivative is recorded[15], but these are very sensitive to base-line drift, and best results are usually obtained by graphical integration of the initial curves.

The line width of the absorption line can be obtained directly from the recorder trace, or the oscilloscope screen. It should be noted that the separation between the peaks of a derivative curve gives the width between points of maximum slope, and this is not, in general, equal to the width at half-power level, which is that normally measured from an oscilloscope trace. It is therefore wise to state which width has been measured when quoting results. The base lines of both the recorder trace and the oscilloscope screen can be calibrated directly by proton-resonance measurements, and these determinations can be made at the same time as the spectra are recorded and injected as marker pips on the trace. A very simple, although less accurate, alternative is to use a mixture of 1, 1-diphenyl-2-picryl hydrazyl and ultramarine as a calibrating sample. Ultramarine is one of the few free radicals with a g-value somewhat displaced from 2·0, and the Δg between it and hydrazyl of 0·025 gives a line separation of 43 gauss at 3·2 cm wavelength. The recorded splitting between these two absorption lines can therefore be measured and equated to 43 gauss.

3.6 ULTIMATE SENSITIVITY

The question is often raised as to what is the ultimate sensitivity of free radical detection. The answer to this depends on how stable the apparatus is and how long-lived the free radicals are. Given a perfectly stable apparatus and a completely stationary free-radical concentration the minimum detectable susceptibility, or number of radicals, can be reduced indefinitely by decreasing the bandwidth, $\Delta \nu$, of equation 3.27 and taking a longer and longer time to sweep through and record the absorption.

Unfortunately the applications requiring highest sensitivity are often those in which fast rates of change in radical concentrations need to be studied. It should be pointed out here that there is an absolute limit to the sensitivity in such cases, set by a fundamental theorem of information theory. This states that the maximum information that can be obtained about a system (or its sensitivity in this context) is proportional to the time taken in acquiring it. The quantitative figures for the case of an X-band electron resonance spectrometer, operating at room temperature, can be obtained from equation 3.25, and the typical data quoted there.

$$\text{I.e. } N_{0_{\min.}} = 3.10^{10}.(\varDelta\nu)^{\frac{1}{2}} = \frac{3.10^{10}}{\tau^{\frac{1}{2}}} \qquad \ldots(3.41)$$

for an ideal noiseless detector and an absorption line width of 1 gauss. τ is the time constant of the detecting system, and thus the minimum time between successive measurements.

Hence if the variation in free-radical concentration is only required at ten-second intervals, a minimum concentration of 10^9 can, in principle, be detected; whereas if the concentration is required at millisecond intervals, the detectable minimum rises to 10^{11}. Since the maximum mass of sample that can be inserted into an X-band spectrometer is of the order of 1 gramme, the above figures can be taken as either the actual minimum number of radicals that can be detected, or the minimum concentration in free radicals per gramme.

It has been seen in the preceding sections that the noise of the detecting systems decreases this sensitivity by a factor of five to ten, and the actual maximum sensitivities at present available at room temperature are

$$N_{0_{\min.}} = (10^{11} \text{ to } 10^{12}).\frac{\varDelta H}{\tau^{\frac{1}{2}}} \qquad \ldots(3.42)$$

where $\varDelta H$ is the line width in gauss.

Improvement in this figure can be expected from only two directions.

(i) Better microwave detectors, such as solid-state masers, or travelling wave tubes; these will increase the sensitivity by a maximum factor of 5.

(ii) The development of more powerful and stable oscillators at shorter wavelengths, with larger magnets. These should decrease the minimum detectable number of free radicals, if present in a small specimen, but will not decrease the minimum detectable radical concentration in a sample of large volume to the same extent.

It may be noted that the sensitivity can be appreciably increased by cooling the sample to very low temperatures, but this often destroys the typical characteristics of the free radical system under investigation.

REFERENCES

[1] FEHER, G. Bell. Syst. Tech. J. 36 (1957) 449
[2] TORREY, H. C. and WHITMER, C. A. Crystal Rectifiers. McGraw-Hill, 1947
[3] NICOLL, G. R. Proc. Instn. elect. Engrs, 101 (1954) 317

REFERENCES

[4] KARPLUS, R. *Phys. Rev.* 73 (1948) 1027

[5] BERINGER, R. and CASTLE, J. G. *Phys. Rev.* 78 (1950) 581

[6] PAKE, G. E., WEISSMAN, S. I. and TOWNSEND, J. *Disc. Faraday Soc.* 19 (1955) 147

[7] BUCKMASTER, H. A. and SCOVIL, H. E. D. *Canad. J. Phys.* 34 (1956) 711

[8] TINKHAM, M. *Proc. roy. Soc. A*, 236 (1956) 535

[9] LLEWELLYN, P. M. *J. sci. Instrum.* 34 (1957) 236

[10] TUCKER, M. J. *Electron. Engng*, 21 (1949) 239

[11] SCHUSTER, N. A. *Rev. sci. Instrum.* 22 (1951) 254

[12] LAWSON, J. L. and UHLENBECK, G. H. E. *Threshold Signals.* McGraw-Hill, 1948, p. 253

[13] RICE, S. O. *Bell. Syst. Tech. J.* 23 (1944) 282; 24 (1945) 46

[14] POUND, R. V. *Rev. sci. Instrum.* 17 (1946) 490

[15] STRANDBERG, M. W. P., TINKHAM, M., SOLT, I. H. and DAVIS, C. F. *Ibid.* 27 (1956) 596

[16] HIRSHON, J. M. and FRANKEL, G. K. *Ibid.* 26 (1955) 34

[17] ROSE-INNES, A. C. *Jour. sci. Inst.* 34 (1957) 276

[18] BERTHET, G. *Onde elect.* 35 (1955) 489

[19] RYTER, C., LACROIX, R. and EXTERMANN, R. *Ibid.* 35 (1955) 490

[20] UEBERSFELD, J. *Ibid.* 35 (1955) 492

[21] GARSTENS, M. A., SINGER, L. S. and RYAN, A. H. *Phys. Rev.* 96 (1954) 53

[22] INGRAM, D. J. E. and TAPLEY, J. G. *Phil. Mag.* 45 (1954) 1221

[23] TOWNSEND, J., WEISSMAN, S. I. and PAKE, G. E. *Phys. Rev.* 89 (1953) 606

[24] POUND, R. V. and KNIGHT, W. D. *Rev. sci. Instrum.* 21 (1950) 219

[25] GINDSBERG, J. and BEERS, Y. *Ibid.* 24 (1953) 632

[26] FEHER, G. *Phys. Rev.* 103 (1956) 500

[27] FEHER, G. *Ibid.* 103 (1956) 834

[28] FEHER, G. *Ibid.* 105 (1957) 1122

[29] OVERHAUSER, A. W. *Ibid.* 92 (1953) 411

[30] CARVER, T. R. and SLICHTER, C. P. *Ibid.* 92 (1953) 212

[31] ABRAGAM, A., COMBRISSON, J. and SOLOMON, I. *Arch. Sci., Genève*, 10 (1957) 240

[32] INGRAM, D. J. E., TAPLEY, J. G., JACKSON, R., BOND, R. L. and MURNAGHAN, A. R. *Nature, Lond.* 174 (1954) 797

[33] BENNETT, J. E., INGRAM, D. J. E. and TAPLEY, J. G. *J. chem. Phys.* 23 (1955) 215

THE THEORY OF HYPERFINE INTERACTION AND LINE WIDTH

4.1 The Parameters of an Electron Resonance Spectrum

There are four important parameters which define any electron resonance spectrum, namely,

(i) the 'g-value' or 'spectroscopic splitting factor' as it is sometimes called

(ii) the value of any splitting of the electronic levels

(iii) the value of any hyperfine splitting or splittings and

(iv) the width of the absorption lines.

In the case of paramagnetic atoms the g-value may vary over a wide range from 1 to 6 or more. This is due to an admixture of the orbital momentum with that of the spin, via the spin-orbit coupling. As a general rule the larger the splitting between the orbital levels of the system the closer the g-value is to 2.0. This explains why the g-values of nearly all free radicals are very close to that of a free-spin, since the high asymmetry produced by the covalent bonds strongly quenches the orbital levels, leaving very little spin-orbit coupling. Even in the case of very symmetrical radicals, such as those derived from the benzene ring, distortion of bond lengths and angles occurs in order to remove the degeneracy, and produces a large orbital splitting. This also accounts for the small spin-lattice interaction and long relaxation times that are encountered in free-radical spectra, in contrast to the very short spin-lattice relaxation times that are normal in electron resonance of paramagnetic atoms. Since the g-values of nearly all free radicals are within $\frac{1}{2}$ per cent of 2·0023, this parameter is not very useful for distinguishing different spectra.

The second parameter listed above, the electronic splitting, only occurs if more than one unpaired electron is present in the atom or molecule. Hence, apart from the case of diradicals or triplet states which are considered separately in a later chapter, no electronic splitting is possible in free radical spectra. The theoretical interpretation has therefore been concerned with the two parameters which do vary in radical spectra, i.e. the hyperfine splitting and line width, and these are now considered separately and in detail.

4.2 HYPERFINE INTERACTION

The existence of well resolved and very characteristic hyperfine structure is one of the great advantages of electron resonance when used to study free radicals. The production of such a hyperfine splitting has already been mentioned in the introductory chapter, and two of the simplest cases were briefly outlined. The general theory is now considered in more detail, and examples quoted in the next chapter, taken from the resonance spectra of various stable radicals, will be found to illustrate the points involved.

The hyperfine splitting of an electron resonance line arises from the interaction between the magnetic moment of the unpaired electron and the magnetic moment of nuclei which are embraced in the molecular orbit of the electron. The general theory of such an interaction for the particular case of one nucleus was given in the early work of ABRAGAM and PRYCE[1] on the electron resonance of paramagnetic atoms and ions. The normal way to express the energy of interaction is in the form of a ' Hamiltonian ', and this may be written as:

$$\mathcal{H}=g_e \cdot g_N \cdot \beta \cdot \beta_N \cdot \left[\frac{S_e \cdot I_N}{|r_e - r_N|^3} - \frac{3\,(S_e \cdot r)\,(I_N \cdot r)}{|r_e - r_N|^5} - \frac{8\pi}{3} \cdot S_e \cdot I_N \cdot \delta(r_e - r_N) \right] \quad \dots (4.1)$$

where
g_e = g-value of the electron
g_N = g-value of the nucleus
β = Bohr magneton
β_N = Nuclear magneton
S_e = electron spin operator
I_N = nuclear spin operator
r_e = vector position of the electron
r_N = vector position of the nucleus
$\delta(r_e - r_N)$ = Dirac delta function for the distance between the electron and the nucleus, normalized in 3 dimensions.

The spin-orbit coupling has been omitted from this expression since the orbital angular momentum is nearly always quenched in the case of free radicals so that the electron behaves as if it had a free spin, with negligible spin-orbit interaction.

This Hamiltonian can be divided into two parts, the first giving an anisotropic hyperfine splitting, while the second gives an isotropic splitting,
I.e.

(i) Anisotropic (Dipolar) Splitting

$$\mathcal{H}^{(1)}=g_e \cdot g_N \cdot \beta \cdot \beta_N \cdot \left[\frac{S_e \cdot I_N}{|r_e - r_N|^3} - \frac{3\,(S_e \cdot r)\,(I_N \cdot r)}{|r_e - r_n|^5} \right] \quad \dots (4.2)$$

103

This can be considered as arising from the classical ' dipole-dipole ' interaction between two magnetic moments.

(ii) Isotropic Splitting

$$\mathscr{H}^{(2)} = g_e \cdot g_N \cdot \beta \cdot \beta_N \cdot \left[-\frac{8\pi}{3} \mathbf{S}_e \cdot \mathbf{I}_n \cdot \delta(\mathbf{r}_e - \mathbf{r}_n) \right] \quad \ldots \ldots (4.3)$$

which is often referred to as the Fermi contact term, and comes from the treatment of relativity theory.

In the paramagnetic salts to which this theory was first applied, the anisotropic hyperfine splitting could be studied in detail by taking measurements on single crystals as they were rotated in the applied magnetic field. Most free radicals are studied in a non-crystalline phase, however, and hence all the anisotropic part of the hyperfine splitting will be smeared out by the random direction. of the radical bonds with respect to the applied magnetic fields In the case of liquids or low-viscosity glasses, where the surrounding molecules are tumbling round the radical in a random manner, it can be shown[2] that the anisotropic hyperfine splitting will in fact average to zero, provided the tumbling frequency is high compared with that of the hyperfine splitting. This condition is nearly always fulfilled for liquid solutions of organic free radicals, and very well resolved spectra can often be obtained from such specimens. The hyperfine splitting is then determined entirely by the isotropic inter-action of equation 4.3.

It will be seen from this equation that the all-important term in the isotropic interaction is the Dirac delta-function $\delta(\mathbf{r}_e - \mathbf{r}_N)$, and this will only have a non-zero value if the molecular orbital occupied by the unpaired electron does not vanish at the position of the nucleus in question—i.e. there must be a finite probability of the electron wave-function at the nucleus if there is to be any hyperfine splitting. In the case of single atoms it is evident that only *s* orbitals will fulfil this condition. Thus *p*, *d* and *f* orbitals will have lobes of electron density pointing out from the origin in different directions, and all have a zero node at the origin, which is the location of the nucleus. The *s* orbitals, however, have no asymmetrical lobes but a density with spherical symmetry, and a finite value at the origin. When passing over to the more delocalized electrons of free radicals, molecular orbitals, instead of atomic orbitals, must be considered, but the same reasoning holds. Thus π orbitals, with one unit of angular momentum, have two lobes pointing in opposite directions, like the atomic *p* orbitals, and also have a zero node at the centre. Only the σ molecular orbitals, corresponding to the atomic s orbitals, will have a finite electron density at the nucleus, and thus the ability to produce an isotropic hyperfine splitting.

4.2.1 Configurational Interaction

From the foregoing discussion it would appear, at first sight, that no aromatic free radical should have any hyperfine splitting. Thus the unpaired electron associated with any system of planar ring structures must be located in the π-orbital system so that it can become delocalized, and the radical stabilized by 'resonance'. The density of the wave-function of the unpaired electron will therefore be confined to regions above or below the plane of the aromatic rings, and will have a zero value at any nucleus in the aromatic structure. If the protons round the edge of an aromatic hydrocarbon are considered in particular, no electron in the π-orbital system will have any interaction with these edge protons. It is found experimentally, however, that aromatic hydrocarbon free radicals[3], such as triphenylmethyl[4] or the naphthalene negative ion[5], have in fact a marked and well-resolved hyperfine splitting.

A situation very similar to this, in which the simple theory predicted no, or very small, hyperfine splitting, whereas a large isotropic splitting was obtained experimentally, occurred in the earlier electron resonance studies on paramagnetic salts. Thus the initial theories of Abragam and Pryce predicted that an ion with a ^6S or $(3d)^5$ ground state, such as Mn^{2+} would have very little hyperfine splitting, whereas BLEANEY and INGRAM[6] showed experimentally that Mn^{2+} had an isotropic splitting of about 100 gauss between successive components. The theory was then developed to account for this by including the effect of configurational interaction. This is the name given to the fact that it is possible for the $(3s)^2 (3d)^5$ ground state to interact with an excited state such as $(3s)^1 (3d)^5 (4s)^1$, and as a result the electron density takes on a small amount of the distribution corresponding to the excited state. It is seen that such an excited state now has unpaired electrons in the s orbitals and hence the possibility of producing a large isotropic hyperfine splitting. The interaction of the nucleus with the s orbitals is, in fact, so large that only a small amount of the excited state need be admixed by this 'configurational interaction' to account for the observed hyperfine splittings.

This concept of configurational interaction can now be extended from the atomic orbitals of a single paramagnetic ion to the molecular orbitals of a complex free radical. In essence the mechanism produces the same result—i.e. some of the σ-orbital character of an excited state is admixed with the pure π-orbital character of the unpaired electron's molecular orbital. This gives a finite density for the unpaired electron's wave-function at the edge protons, or other nuclei, and produces the isotropic hyperfine splitting which is observed experimentally.

The quantitative treatment of configurational interaction, as applied to aromatic free radicals, has been undertaken by various authors, WEISSMAN[7], McCONNELL[8], JARRETT[9] and BERSOHN[10], in particular. In the following summary the treatment given follows that due to Weissman[7], but is very similar, in essence, to that of the other workers.

To make the problem somewhat specific the bonding between a carbon and hydrogen atom is first considered. The bonding orbital between these two is represented by σ_B, and the main configuration of the ground state, in which the unpaired electron producing the free radical properties is in a π orbital, can then be represented by

$$[\text{Filled orbitals}].(\sigma_B)^2.\pi. \qquad \qquad ....(4.4)$$

It may be noted that the σ_B orbital is in fact made up of a carbon sp^2 hybrid orbital and the $1s$ orbital of the proton. It is also possible to form an anti-bonding orbital in the same way but with opposite signs for the overlapping carbon and hydrogen orbitals, and this may be denoted by σ_A. Moreover an excited state of the C–H bond will be formed if one of the two paired electrons in the σ_B orbital is promoted to the σ_A orbital. Thus one of the excited states can be written as

$$[\text{Filled orbitals}].(\sigma_B)^1.\pi.(\sigma_A)^1 \qquad \qquad ....(4.5)$$

This excitation corresponds to the change from $(3s)^2 (3d)^n$ to $(3s)^1$ $(3d)^n (4s)^1$ for the paramagnetic ions.

Other excited states can, of course, be formed by promoting one of the σ_B electrons into a π orbital, and these would also leave an unpaired electron in a σ orbital. It is not possible to admix a $(\sigma_B)^2 \cdot \pi$ configuration with a $(\sigma_B)^1 (\pi)^2$ configuration, however, since one of the conditions of configurational interaction between orbitals is that they should have the same symmetry with respect to reflection in the aromatic plane.

The configurational interaction can therefore be used to admix into the main ground-state electron distribution of equation 4.4, some of the electron distribution of the excited state given by equation 4.5. To place this on a quantitative basis wave-functions must be obtained for the ground and excited states, and a linear combination of these taken, and their distribution at the edge protons then determined. Since all the spins of the filled orbitals will cancel these may be ignored, and only the two three-electron configurations $(\sigma_B)^2.\pi$ and $(\sigma_B)^1.\pi.(\sigma_A)^1$ need be considered. The wave-function of the unperturbed ground state can then be written as

$$\phi_g = A \parallel a.\sigma_B(1), \ \beta.\sigma_B(2), \ a.\pi(3) \parallel \qquad \dots(4.6)$$

and those of the excited state by the three possible spin-orbital configurations

$$\phi_1 = A \parallel a.\sigma_B(1), \ \beta.\sigma_A(2), \ a.\pi(3) \parallel$$

$$\phi_2 = A \parallel \beta.\sigma_B(1), \ a.\sigma_A(2), \ a.\pi(3) \parallel \qquad \dots(4.7)$$

$$\phi_3 = A \parallel a.\sigma_B(1), \ a.\sigma_A(2), \ \beta.\pi(3) \parallel$$

where A represents the antisymmetrization and normalization operator, and a, β represent the two possible spin quantizations. Linear combinations of these three functions must be taken to describe the excited state accurately, since the correct functions must be eigenvalues not only of S_Z (i.e. with $S_Z = \frac{1}{2}$) but also of S^2. The three linear combinations which are eigenfunctions of S^2 are:

$$\phi_x = \frac{1}{\sqrt{2}}(\phi_1 - \phi_2) \qquad \dots(4.8)$$

$$\phi_y = \frac{1}{\sqrt{6}}(\phi_1 + \phi_2 - 2\phi_3) \qquad \dots(4.9)$$

$$\phi_z = \frac{1}{\sqrt{3}}(\phi_1 + \phi_2 + \phi_3) \qquad \dots(4.10)$$

and the first two have an eigenvalue of $\frac{3}{4} \times \left(\frac{h}{2\pi}\right)^2$ while the last has one of $\frac{15}{4} \times \left(\frac{h}{2\pi}\right)^2$.

The first two also correspond to a doublet state while the last corresponds to a quartet state, which is of different spin multiplicity to the ground state and therefore cannot admix with it. Thus configurational interaction can only admix some of ϕ_x and ϕ_y components of the excited state, and the configurational admixture in the ground state can therefore be written as

$$\phi_0 = \phi_g + \lambda_x.\phi_x + \lambda_y.\phi_y \qquad \dots(4.11)$$

provided λ_x and $\lambda_y \ll 1$.

The hyperfine splitting will be produced by the last two terms. In fact only one of these is effective since ϕ_x corresponds to a combination of the singlet (σ_B) (σ_A) with π and hence has paired spins in the σ orbitals. The quantitative calculation therefore resolves into a determination of λ_y.

It can be shown by first-order perturbation theory that

$$\lambda_y = \frac{\text{The expectation value of } \mathscr{H}^\star}{W_y - W_0} \qquad \ldots (4.12)$$

where $\mathscr{H}^\star$ is the Hamiltonian for the repulsion energy terms of the three electrons.

If it is assumed that the π orbital is a linear combination of the carbon $2p_z$ orbitals, and that the terms originating from other carbon atoms can be neglected because they only give minor corrections to λ_y, the expression for λ_y may be written as

$$\lambda_y = \rho_i \cdot \frac{\sqrt{6}}{2}\left[\frac{\int \sigma_A(1) \cdot \phi_i(2)\left|\dfrac{e^2}{r_{12}}\right|\phi_i(1) \cdot \sigma_B(2) \cdot d\tau}{W_y - W_0}\right] \qquad \ldots (4.13)$$

where ϕ_i represents the $2p_z$ orbital of the carbon atom labelled i, and ρ_i is the unpaired spin density on that carbon atom. The integral is often termed the 'electrostatic repulsion integral', and denoted by G, since it corresponds to the interaction of the repulsion term $\dfrac{e^2}{r_{12}}$, between the bonding and antibonding orbitals. Equation 4.13 can therefore be written as

$$\lambda_y = -\frac{1}{2}\sqrt{6} \cdot \frac{G}{\varDelta W} \cdot \rho_i \qquad \ldots (4.14)$$

The final stage in the calculation of the actual hyperfine splitting is to relate the hyperfine interaction energy to the value of λ_y, derived above.

In the presence of a strong magnetic field applied along the z axis, the isotropic hyperfine interaction as expressed in equation 4.3, can be written as

$$\mathscr{H}^{(2)} = -\frac{8\pi}{3} \cdot \beta \cdot \beta_N \cdot \sigma_{ze} \cdot \sigma_{zN} \cdot \delta(\mathbf{r}_e - \mathbf{r}_N) \qquad \ldots (4.15)$$

where σ_{ze} and σ_{zN} are the spin operators for the z component of the electron and nuclear spin respectively. The first-order hyperfine

108

interaction energy is then given by the expectation value of $\mathcal{H}^{(2)}$, which is

$$\Delta E = \frac{32\pi}{3\sqrt{6}}\lambda_y \cdot \mu_N \cdot \mu_B \cdot \sigma_B(\mathbf{r_N})\sigma_A(\mathbf{r_N}) \qquad \dots (4.16)$$

There are several general features of this result which are of considerable importance.

(1) The hyperfine splitting is *linearly* dependent on λ_y. Since, from equation 4.14, λ_y is linearly dependent on ρ_i, it follows that the hyperfine splitting produced by a proton attached to a given carbon atom is *linearly* proportional to the unpaired electron density on that carbon atom. This is a result which is of great value in the analysis of radical spectra. If it is assumed that the proportionality factor between λ_y and ρ_i is constant for all C–H bonds in an aromatic hydrocarbon, the relative densities of the unpaired electron at different carbon atoms can also be calculated directly from the hyperfine splittings. I.e., the hyperfine splitting due to one proton then reduces to

$$\Delta H = Q \cdot \rho_i \qquad \dots (4.17)$$

where Q is a constant for all aromatic hydrocarbons.

(2) The hyperfine interaction depends on the product of the probability amplitude of the state from which the electron is excited (σ_B) and that of the state into which it is excited (σ_A). Both σ_B and σ_A are non-vanishing at the carbon nucleus as well as the hydrogen nucleus, and it therefore follows that if there is any C^{13} in the aromatic ring this will also produce a hyperfine splitting, as has been observed experimentally[11]. It is therefore evident that configurational interaction will not only allow the unpaired electron in the π-orbital system to interact with edge protons, but also with the nuclei in the aromatic ring structure itself.

This theory of configurational interaction has been applied in some detail to various aromatic free radicals and can account for the precise splittings of the spectra in a large number of cases. Examples of this kind of interpretation are given in the next chapter when the aromatic radicals are considered in detail.

From the quantitative point of view it is interesting to note that, on a simple first order approach, this theory predicts that the overall proton hyperfine splitting for all aromatic hydrocarbon free radicals should be approximately constant and equal to 28 gauss. This result was first derived by JARRETT[9], who carried McConnell's[8]

109

work through in more detail, and derived an expression very similar to that of Weissman's above. The actual hyperfine splitting that he obtained, by considering the configurational interaction between the two doublet states in conjunction with the Fermi isotropic hyperfine interaction, is given by

$$\Delta H = \frac{16\pi}{3} \cdot \beta_N \cdot \beta \cdot \frac{[h^2(r_H) - \sigma_c^2(r_H)]}{1 - S_{\sigma h}^2} \cdot \lambda \qquad \ldots\ldots(4.18)$$

where $h^2\,(r_H)$ and $\sigma^2\,(r_H)$ represent the squares of the magnitude of the hydrogen's electron wave function and the carbon σ wave function at the position of the proton, and $S_{\sigma h}$ is the overlap integral. The part involving $h^2\,(r_H)$ can be taken as the hyperfine splitting of a free hydrogen atom equal to 510 gauss. The σ-bond contribution can be determined from self-consistent field configurations, which give $\sigma^2\,(r_H)/h^2\,(r_H) = 0{\cdot}17$, whence

$$\Delta H = 423.\lambda/(1 - S_{\sigma h}^2) \text{ gauss} \qquad \ldots\ldots(4.19)$$

Values of λ and $S_{\sigma h}$ were then calculated by Jarrett by the use of Slater wave-functions and the exchange and Coulomb integrals following the method of ALTMANN[12]. These gave values of $\lambda = 0{\cdot}024$ and $S_{\sigma h} = 0{\cdot}80$ for typical C–H bond distances, which yield

$$\Delta H = 28 \text{ gauss} \qquad \ldots\ldots(4.20)$$
(i.e. Q of equation 4.17 = 28 gauss)

It has been assumed throughout this calculation that the unpaired electron spends all its time on the one carbon atom, and if the odd-electron density on the carbon atom is less than unity the splitting due to the attached proton will be correspondingly less than 28 gauss. On the other hand, if the unpaired electron of an aromatic radical is spread uniformly around the outer carbon atoms, it follows that the width of the whole hyperfine pattern should still be 28 gauss, although the pattern itself may be considerably complicated by interaction with several protons.

This first attempt at quantitative calculation of configurational interaction in aromatic free-radicals gives surprisingly good agreement with the experimental results summarized in *Table 4.1*, where the splitting between extreme hyperfine lines is listed for various aromatic radicals. It is seen that the values are all remarkably close to 28 gauss.

There are some radical spectra, however, that of perinaphthene[13] for example, in which this extreme splitting is considerably greater than 28 gauss, and it was at first thought that this was evidence that the treatment of configurational interaction outlined above was not generally applicable. A more rigorous analysis[14] showed, however, that the possibility of ' negative spin density ' at certain

Table 4.1.   Extreme Hyperfine Splitting of Aromatic Free Radicals

(i) *Negative Ions*	Hfs. Splitting gauss
Diphenyl	24
Naphthalene	29
Anthracene	28
Perylene	30
Tetracene	27
Phenanthrene	25
Pyrene	32
Coronene	23
(ii) *Positive Ions*	
Anthracene	33
Tetracene	30
Perylene	33

carbon atoms had been ignored, and that if this were included in the theory, splittings of greater than 28 gauss could be obtained. This ' negative spin density ' is produced by the disturbing effect of the unpaired electron on the orbitals of the paired electrons. This may be viewed as a form of spin repulsion with a result that there is an effective partial unpairing of the previously balanced orbits, and an extra odd electron density is thus produced of opposite direction to that of the unpaired electron responsible for it. This new spin density is therefore given a negative sign, and the positive spin densities also increase in magnitude so that the algebraic sum of the spin densities remains unity. The actual hyperfine splitting is independent of the sign of the spin density, however, and the extreme splitting is therefore proportional to the sum of the moduli of the spin densities, which can be greater than unity.

This introduction of the ' negative spin density ' into the configurational interaction has been treated quantitatively by McCONNELL[14], and can be expressed as follows. The hyperfine splitting due to an aromatic proton is still represented by

$$\Delta H = Q . \rho_i$$

with $Q = 28$ gauss, but ρ_i is taken to be the expectation value of the spin density operator $\boldsymbol{\rho}_i$ where

$$\boldsymbol{\rho}_i \sum_k S_{kz} = \sum_k \Delta_i(k) . S_{kz} \qquad \dots . (4.21)$$

111

$\Delta_i (k)$ is the 'atomic orbital delta function', i.e. a three-dimensional step function such that $\Delta_i (k)=1$ when the electron k is in a p_z orbital centred on the ith carbon atom, and is zero elsewhere. S_{kz} is the component of the spin of electron k in the strong field direction.

The essential point is that although $\Sigma_i \rho_i = 2S$ where S is the total spin, ρ_i can be positive or negative. The total spread of the resonance hyperfine spectrum will be $\Sigma_i |\Delta H_i|$ and equal to $Q.\Sigma_i |\rho_i|$, and it is now possible for the summation to be greater than unity.

The effect of 'negative spin densities' can probably be best illustrated by a specific example, and the case of the allyl alcohol radical may be so considered. This has the formula

It is not an aromatic compound but the binding of a carbon atom to another carbon and two protons is somewhat similar, in a first approximation, to that of a carbon in an aromatic ring. In the first case the three sp^2 lobes of the carbon hybrid orbital overlap with two $1s$ orbitals of the attached protons, and one lobe of the adjacent carbon atom; whereas in the second case one of the hydrogen $1s$ orbitals is replaced by another sp^2 lobe of the other adjoining carbon atom. Therefore a similar splitting parameter, Q, will be expected for all planar groups. The best empirical value for Q is 23 gauss derived from the spectra of the benzene negative ion, and this is confirmed by the fact that the planar CH_3 radical has a splitting parameter of 22·5 gauss. The empirical value of 23 gauss rather than the theoretical 28 gauss will therefore be used in the calculations below. Thus, a spin density of ρ_i on the carbon atom of an aromatic will produce a doublet with a splitting of $23.\rho_i$ gauss for an attached proton, and the same spin density ρ_i on the end carbon atom of an allyl, or similar, radical will produce a triplet with a splitting between successive lines of $23.\rho_i$ gauss.

There are thus two steps necessary in any detailed comparison of theory with experiment. The first is to calculate the value of the splitting parameter, Q, produced by configurational interaction (as carried out by McConnell and Jarrett for the aromatic rings where $Q=28$ gauss). The second step is to calculate the spin density ρ_i at each carbon atom, by molecular orbital or valence bond theory. The hyperfine splitting for protons attached to the ith carbon atom is then given by $\rho_i \cdot Q$. The calculation of spin densities can now be performed with high accuracy, and the work on the allyl radical affords a good example of this.

In the simple molecular orbital picture of an allyl radical there will be two paired electrons in the bonding ground state orbital which has an electron density distributed between all three carbon atoms. The unpaired electron will be in the antibonding orbital, however, which has opposite sign at either end of the carbon chain and hence a zero node at the middle. The odd-electron density is therefore expected to be $\frac{1}{2}$ on each of the end carbon atoms, and zero at the centre carbon. The configurational interaction will therefore produce a splitting of $\frac{1}{2} Q$ gauss for each proton spin. Since there is a total of three equally-coupled protons for the two end carbon atoms it follows from *Figure 7(b)*, that four hyperfine components are to be expected, with an intensity ratio of $1:3:3:1$, and a splitting between each of $\frac{1}{2} Q$ gauss.

Exactly the same structure is predicted by the simple valence-bond picture in which a resonance between the two structures for the carbon chain of

$$\uparrow C \; - \; C \; = \; C$$

and $$C \; = \; C \; - \; C\uparrow$$

is envisaged. This also results in a spin density of $\frac{1}{2}$ for each of the end carbon atoms, and of zero for the central atom.

If the disturbing effect of the unpaired electron on the other orbitals is now considered, however, it is found, from a first order calculation, that a 'negative spin density' of $-\frac{1}{3}$ is produced on the central carbon atom, and of $+\frac{2}{3}$ on each of the outer carbon atoms. This will give a markedly different electron resonance hyperfine pattern from that predicted by the simple theory. The splitting between the four main lines will be increased from $\frac{1}{2} Q$ to $\frac{2}{3} Q$ gauss, while each line is also split into two sub-components of $\frac{1}{3} Q$ gauss separation, by the interaction with the proton on the central carbon atom.

The detailed calculation, taking into account all possible π electron configurations has in fact been carried out by LEFKOVITS, FAIN and MATSEN[15] for this molecule, and this gives for the accurate spin densities on the three carbon atoms: $0\cdot622$; $-0\cdot231$; and $0\cdot622$. The predicted hyperfine pattern therefore has a splitting of about 14 gauss between the positions of the four main components, and these are each split into two subcomponents with a separation of 5 gauss. The same calculation can be applied to the allyl alcohol radical, though the oxygen atom here may withdraw some of the spin density if the frequency of rotation of the OH group is not greater than the microwave frequency. The experimental results[16] in this case gave a splitting of 12 gauss between each of the four hyperfine components, and the resolution of the spectrum in the solid phase was not sufficient to separate the small doublet splitting of each line.

It is therefore evident that the theory of configurational interaction can be applied to a large number of radical structures, and that detailed calculations, in which the effect of 'negative spin densities' is also included, predict hyperfine patterns which are in very good agreement with the experimental results.

In concluding this subject, the main points of the theory of configurational interaction may be summarized as follows:

(i) The unpaired electron of a free radical occupies a π molecular orbital, and is thus stabilized by delocalization or resonance. Such an orbital has zero density at the position of nuclei in the plane of an aromatic ring, however, and hence no hyperfine splitting of the spectrum is to be expected.

(ii) In order to explain the observed isotropic hyperfine splittings it is necessary to assume that the π orbital has some of the σ type orbital admixed into it, and a finite odd-electron density is therefore produced at the edge nuclei.

(iii) This admixture of the orbitals is brought about by configurational interaction in which the ground state $(\sigma_B)^2.\pi$ has some of the excited state $(\sigma_B).\pi.(\sigma_A)$ admixed with it by the mutual electron repulsions.

(iv) This configurational interaction can be treated quantitatively and it can be shown that the resultant hyperfine splitting is linearly proportional to the density of the unpaired electron on the carbon atom concerned. In the particular case of an aromatic C–H bond, substitution of quantitative values shows that the splitting produced by the interaction of one proton with an unpaired electron density of unity on the carbon atom is equal to 28 gauss.

(v) In the more refined calculations of molecular orbital, or valence bond, theory it is found that the concept of 'negative spin density' also occurs. This modifies the maximum splitting of 28 gauss, predicted for aromatic species, since the

114

splitting is independent of the sign of the spin density and the sum of the moduli of the spin densities can now be greater than unity.

(*vi*) Very good agreement with the experimental hyperfine pattern can usually be obtained if the fractional unpaired spin density, ρ_i, on each carbon atom is calculated, using as refined a molecular orbital or valence bond treatment as possible; and the splitting due to each proton attached to ρ_i is then put equal to $Q'.\rho_i$ gauss. This type of calculation is illustrated very well by the spectra of the aromatic radicals considered in the next chapter.

4.2.2 *Hyperconjugation*

In the preceding section it has been shown how configurational interaction between electron orbitals results in hyperfine structure from the edge protons of an aromatic radical. This does not explain, however, how a hyperfine splitting can be produced by groups such as methyl attached to the ring. As a specific example the case of tetra-methylated benzo-semiquinone can be taken, which has the formula

The carbon atoms of the CH_3 groups now have four bonds, and hence have sp^3 configurations, with lobes overlapping the three $1s$ orbitals of the protons and one sp^2 lobe of a ring carbon. There are thus no free p orbitals to take part in the π-bond system of the ring and therefore no unpaired electron density is to be expected at the methyl groups. No direct configurational interaction with the protons, of the type considered in the last section, is possible in this case.

It is found, experimentally, however, that radicals containing such methyl groups do have hyperfine splittings of considerable magnitude, and there must therefore be some other mechanism whereby the unpaired electron can move out from the ring system on to the edge protons. This mechanism can be explained by considering

115

the orbitals associated with one of the CH_3 groups attached to the ring. These can be illustrated thus:

Side view   End view

In this diagram the $2p_z$ orbital of the ring carbon, and the $1s$ orbitals of the hydrogen atoms are shown as dotted curves. If the $1s$ orbital of the lower proton, H_b, is taken with a negative sign, then it is evident that the combination of $(h_a - h_b)$, where h_a and h_b are the wave functions of the H_A and H_B protons, will have the same symmetry as the p_z orbital of the carbon, as drawn. There will thus be a possibility of direct overlap between the $2p_z$ orbital of the ring carbon atom and a linear combination of two of the proton orbitals. Since the p_z orbital forms part of the π-orbital system containing the unpaired electron, it is evident that this produces a very direct mechanism whereby the odd-electron density can be shared with the protons. The interaction has so far been confined to two protons, but the three protons of the methyl groups will in fact be rotating at high speed round the direction of the C–C axis, and hence all three will, in turn, participate in this overlap, and an equal hyperfine splitting from all three will be obtained. On an alternative view[17] the overlap can be regarded as involving all three proton orbitals at once. Thus if the protons are considered as rotated by 30° in a clockwise direction from their positions in the above diagram, a wave function with symmetry approximating to that of the p_z orbital can be obtained from $[h_a - \frac{1}{2}(h_b + h_c)]$.

This type of interaction is termed 'hyperconjugation' and it is to be noted that it depends on a spatial overlap of wave-functions with the same symmetry or pseudo-symmetry, and not on the mutual interaction of electrons involving excited levels, as in configurational interaction. It is therefore a more direct and potent effect than the latter, and very nearly the same hyperfine splitting is produced by each proton of the methyl group, as for a proton connected directly to the ring carbon atom. I.e. $\Delta H \approx 30. \rho_i$ for each methyl proton where ρ_i is the unpaired spin density on the *ring* carbon atom.

Methyl groups are not the only configurations for which such a coupling is possible, two protons attached directly to a ring carbon atom form another case in which a suitable linear combination of their orbitals will produce a wave-function with symmetry similar to that of the π system.

It is therefore evident that interactions do exist in the molecular orbital systems whereby the unpaired electron density can enter the $1s$ orbital of the protons, and produce an isotropic hyperfine splitting. This isotropic splitting, due to the Fermi contact term in the Hamiltonian, is by far the most important in free-radical studies, since the anisotropic terms are usually either averaged to zero by molecular motion, or only appear as an extra broadening.

4.3 Anisotropic Hyperfine Splitting—Dipolar Interaction

In the preceding section only the hyperfine interaction arising from the Fermi contact term has been considered. This will always be present, and since the anisotropic dipolar interaction is often averaged to zero in liquids, the isotropic term is often the only interaction to be considered. In solid or viscous phases, however, the anisotropic dipolar interaction may be considerable and, even if smeared out by a random orientation of the molecules, it may contribute very appreciably to the width of the isotropic hyperfine lines.

This anisotropic interaction is similar to that between two classical magnetic dipole moments, and has been given in full vector form in equation 4.2. If this is re-written with scalar quantities the expression becomes

$$\mathscr{H} = g_e \cdot g_N \cdot \beta \cdot \beta_N \sum_k \frac{(3\cos^2\theta_k - 1)}{r_k^3} \qquad \ldots (4.22)$$

or
$$\Delta H = B \cdot \sum_k \frac{(3\cos^2\theta_k - 1)}{r_k^3} \text{ gauss} \qquad \ldots (4.23)$$

where B is equal to 28 for proton interaction and r_k is the distance between the nucleus and the unpaired electron, in Å, averaged over the wave-functions concerned. θ_k is the angle between the direction of the applied magnetic field and the line joining the electron and nucleus.

It may be noted that the above expression for ΔH can be derived directly from classical theory. Thus the magnetic moment of the nucleus can be considered as a permanent magnet of moment $M_I\, g_N\, \beta_N$, along the direction of the applied magnetic field. Its moment at right angles to this will average to zero. The component of field in the direction of the dipole axis, z, at a point at distance, r, is known to be

$$Hz = \mu \cdot \frac{3z^2 - r^2}{r^5} = \mu \frac{3\cos^2\theta - 1}{r^3}$$

where $\cos \theta = z/r$, and is the angle between the z axis and the line joining the dipole to the point concerned. If the electron is now considered to be at this point, and the dipole moment is that given above for the nucleus, the extra field produced will therefore be

$$\Delta H = M_I g_N \beta_N \left(\frac{3 \cos^2 \theta - 1}{r^3} \right) \qquad (\ldots.4.23a)$$

This equation is in fact identical to 4.23, since the conversion from the interaction energy of the Hamiltonian of 4.22 to a field splitting requires a division by $g_e.\beta$. The value of $1/r^3$ can be taken, to a first approximation, as the average over a hydrogen-like wave function.

In most free radical studies the specimen is in a non-crystalline form. Hence over a cross-section of molecules the direction of any particular bond will be quite random with respect to that of the magnetic field. The splitting produced by interaction with any particular proton will therefore have a continuous range of values within the limits of the $(3 \cos^2 \theta - 1)$ variation. This function has a maximum of 2 when θ equals zero, and of 1 when θ equals $\frac{\pi}{2}$. Since there is only one direction corresponding to θ equals zero, but a whole ' plane ' of directions corresponding to θ equals $\frac{\pi}{2}$, it follows that more molecules will have a splitting approaching the value of B/r^3 than the value of $\frac{2.B}{r^3}$. The net result of super-imposing all the different values of splitting[18] is therefore to give a curve with two large peaks of separation B/r^3 gauss, and two subsidiary shoulders with a splitting of twice this value, as shown in *Figure 34(a)*. All other forms of broadening have been ignored in drawing this curve, and when these are included the resultant line is more like that of *Figure 34(b)*.

The dipolar interaction has so far only been treated as between the electron and one proton, whereas in most free-radicals there are several interacting protons. The complete summation over all of these, as indicated by equations 4.22 and 4.23, must then be performed. This problem is very similar to that encountered in calculating the line shapes of nuclear absorption lines in solids when there is more than one interacting nucleus. It has been treated in quantitative detail for some specific cases by ANDREW and BERSOHN[19] and others[20], but, as a general rule, all details of resolved component lines are lost when the other broadening processes are included, if more than one interacting proton is present. This remark is even more applicable in the case of electron resonance, and it is only possible to obtain resolved component lines from dipolar interaction in an amorphous sample if the interaction is predominantly with a single proton. Such a spectrum may be obtained if the unpaired electron becomes partially localized on an oxygen atom which has a proton held close to it for steric reasons[21].

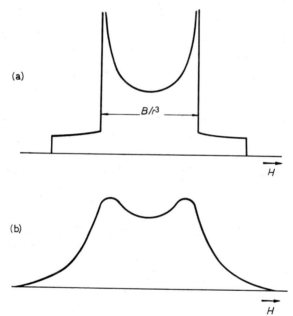

Figure 34. *Splitting produced by dipolar interaction with one proton. (a) With no other broadening present; (b) With broadening due to other sources included*

The magnitude of the splitting expected from this type of interaction can be seen from calculations on a typical C–H bond of length $1 \cdot 07$ Å. This gives a value for ΔH of 23 gauss from equation 4.23, and it will be seen that such a value is larger than most individual isotropic hyperfine splittings. In a large number of cases the dipolar interation is considerably reduced, however, by motional narrowing processes. These are considered in more detail under the heading of ' line widths ' in the next section, where it is seen that a typical width for an absorption line from a free radical trapped in a hydrocarbon glass is about 15 gauss, and most of this is attributed to ' partially narrowed ' dipolar hyperfine interaction.

The effect of anisotropic dipolar interaction on the hyperfine structure of most free radicals can therefore be summarized as follows:

(*i*) The hyperfine splitting will vary as $(3 \cos^2 \theta - 1)$, and hence for any specimen other than a single crystal, will have a ' smeared out ' shape similar to *Figure 34*.

(*ii*) Only if the interaction is predominantly with one proton will any resolved pattern be obtained. This will appear similar to *Figure 34(b)* and the splitting between the two central peaks determines the distance between the proton and the atom localizing the unpaired electron, as in equation 4.23.

(*iii*) The above considerations apply only to molecules which have all their atoms held rigidly. If internal motion such as rotation or jumping is present, the splitting will be reduced, and appear mainly as an increased width for the isotropic hyperfine components. The limit is reached in the liquid phase when the rapid tumbling motion averages the dipolar interaction to zero[2].

4.4   LINE WIDTHS

The interactions which affect the widths of electron resonance absorption lines can be grouped under three general headings

(*i*) Dipolar spin-spin interaction
(*ii*) Spin-lattice interaction, and
(*iii*) Exchange interaction.

The case of dipolar spin-spin interaction has already been introduced as responsible for the anisotropic hyperfine splitting considered in the last section, and it was seen there that a smeared out anisotropic hyperfine splitting was, in effect, another form of line broadening. Dipolar spin-spin interaction can also occur between the electron spins, however, and if the free radical concentration is high enough this will produce a much larger broadening than that due to the nuclear interaction.

The spin-lattice interaction is the term given to any process whereby the spins lose their extra energy to the molecule or solid as a whole, instead of sharing it with other spins, as in spin-spin interaction. There are several different processes that may contribute to this interaction, but most act via the spin-orbit coupling, since the thermal vibrations of the molecule or ' lattice ' are more directly coupled to the orbital motion of the electrons.

120

Both the spin-spin and spin-lattice interactions can be characterized by 'relaxation times', defined as the time taken for the spin
system to lose $\frac{1^{th}}{e}$ of its excess energy. The spin-lattice relaxation
time, so defined, is denoted by T_1, and in early electron resonance
work on paramagnetic atoms, the spin-spin relaxation time was
similarly defined as T_2. If these are the only two interactions to
be considered the line width may be written as

$$\Delta\omega = \frac{1}{T_1} + \frac{1}{T_2} \qquad \dots (4.24)$$

In practice the temperature of the paramagnetic samples was
always reduced until the spin-lattice interaction was negligible, so
that the line width was determined entirely by the spin-spin interaction and

$$\Delta\omega = \frac{1}{T_2}$$

or $g(\omega_0)_{max.} = T_2/\pi \qquad \dots (4.25)$

where $g(\omega - \omega_0)$ is the normalized line shape function and has a
maximum at ω_0, as defined in equation 2.12 and represented in
Figure 35(c).

In more recent work equation 4.25 has been taken as the definition
of T_2, which is now called the 'transverse relaxation' time following the terms used in nuclear resonance. In a large number of
cases these two definitions of T_2 give the same value, but this is

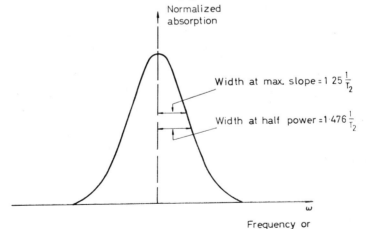

Figure 35. (a) Gaussian line shape

121

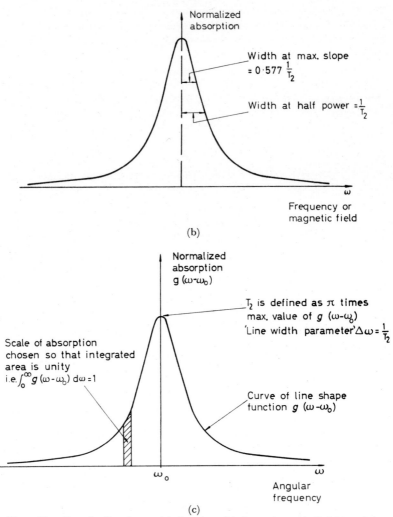

Figure 35. Absorption line shapes. (a) Gaussian; (b) Lorentzian; (c) Definition of the line shape function (see p. 128 for equations of line shapes)

not always so. Since saturation effects and comparatively long values of T_1 are obtained in electron resonance of free-radicals, and the position is thus somewhat similar to nuclear resonance conditions, the definition of T_2 implied by equation 4.25 will be used throughout this book.

It should be noted that strong interactions give rise to short relaxation times and thus to wide lines, and if T_1 is less than 10^{-7} sec the spin-lattice interaction will broaden the electron resonance line. Such strong interactions are common in paramagnetic atoms but very rare in free-radicals, where there is much smaller spin-orbit coupling. Free-radical spectra are therefore very seldom broadened by strong spin-lattice interaction, but may in fact quite easily be broadened by too weak a spin-lattice interaction. This will produce a 'saturation broadening' at high power levels, as is discussed in a following section. In this way the effect of T_1 on free radical spectra is more similar to that in nuclear resonance than that in electron resonance of single paramagnetic atoms.

The other major interaction affecting the line widths of electron resonance spectra is 'exchange'. This arises from the fact that it is possible for electrons to exchange their spin orientation between different atomic or molecular orbitals. If the exchange is between electrons in identical energy states, as is nearly always the case with free radicals, then the effect is to average out the dipolar broadening and narrow the line width. There is thus a very close similarity between the narrowing produced by the rapid tumbling of molecules in a liquid, and the exchange narrowing that can occur in a solid. In the former case the *position* of the interacting nuclear dipoles is modulated and their effect averaged out by Brownian motion. In the latter case the *orientation* of the electron magnetic dipoles is modulated and their broadening reduced. In both cases a new mechanism has been introduced whereby magnetic energy can be converted into thermal energy, and a new value for the spin-lattice relaxation time is thus obtained.

These three interactions will now be considered separately, and in somewhat more detail, together with 'saturation' effects that can occur when T_1 has a large value.

4.4.1 *Spin-spin or Dipolar Broadening*

Spin-spin interaction can take place between the unpaired electron and nuclear spins in neighbouring atoms, and also between the unpaired electron and the spins of unpaired electrons of other molecules. The latter interaction is by far the stronger since the magnetic moment of the electron is about a thousand times that of the nucleus, but since the nearest unpaired electrons are usually much farther away than the nearest nuclei, the actual splitting or broadening produced by the two interactions is often of the same order. The type of interaction is of the same nature for each case, and is given quantitatively in equations 4.22 and 4.23. In the case of interaction between two unpaired electrons the nuclear g factor and magneton of these equations are replaced by the electron g-value and Bohr magneton. In both cases a $(3 \cos^2 \theta - 1)$ variation of splitting with angle is obtained, and if studies are made on single crystals, this splitting produced by nearest neighbours can be resolved and its variation with change of angle confirmed, for both the nuclear[22] and electronic[23] interactions. In the case of amorphous compounds, and hence for most free radicals, only the total splitting integrated over all directions is obtained, as in *Figure 34*, and the net effect of electronic or nuclear spin-spin interaction is thus to produce a line broadening, instead of a line splitting.

The theory of spin-spin interaction has been treated by VAN VLECK[24] and by PRYCE and STEVENS[25], and the expression for the mean square width of the line can be written for the electronic case as

$$(\Delta H)^2 = \tfrac{3}{4} S(S+1) g^2 . \beta^2 . \sum \left(\frac{1 - 3 \cos^2 \theta}{r^3} \right)^2 \qquad \ldots . (4.26)$$

A second term in this equation, expressing the interaction with dissimilar dipoles of different g-values, has been omitted since this case rarely occurs in free-radical studies. Two separate effects now contribute to the expression of 4.26, the first is the actual component of magnetic field produced by neighbouring electrons at the radical in question, and the second is the fact that the rotating components of the neighbouring spin precessions will couple and reduce the normal life-time of the energy state. It can be seen that the only parameter in this expression is the distance between the spins, r. It therefore follows that the only way to reduce electronic spin-spin interaction is to reduce the concentration of the free radicals until other forms of broadening predominate. In most free-radical studies the concentration is small and this form of broadening is therefore usually not important. The dipolar spin-spin interaction with the nearest magnetic nuclei often produces a larger broadening, unless it has been averaged out in a liquid sample by the Brownian tumbling motion[2].

The dipolar spin-spin interaction with the surrounding nuclei has already been discussed at some length in section 4.3 and quantitative expressions for its effect are given there. It was seen that there are two extreme cases, depending on the state of the surrounding medium. If all the molecules and nuclei within them are held rigidly in position then full dipolar interaction averaged over a $(3 \cos^2 \theta - 1)$ splitting will occur. This produces a line shape as illustrated in *Figure 34* for interaction with one proton, and an even less resolved pattern when the electron is interacting with several protons. This broadening will be of the order of 25 gauss for C–H bonds of about 1 Å length and may well obscure most of the detail of the isotropic hyperfine pattern. The other extreme case is when the molecules are tumbling in a random manner around the free-radical so that the frequency at which protons change position is greater than the frequency corresponding to the hyperfine splitting. It has been seen[2] that the dipolar interaction is averaged to zero by this process and no broadening of the absorption lines is produced.

It is evident that a whole range of intermediate conditions will be found between these extremes, when the rate of the 'tumbling motion' is of the same order as the frequency of the hyperfine splitting. A quantitative treatment of the general case has been given by McCONNELL[26] and an expression for the resultant line width similar to that derived in his work can be given as:

$$\Delta\omega \approx \frac{1}{T_2'} + \frac{1}{T_1}$$

where
$$\left(\frac{1}{T_2'}\right)^2 = \frac{2\pi}{h^2}b^2{}_{lk}\ \tan^{-1}\frac{2.\tau_c}{T_2'}$$

and
$$\frac{1}{T_1} = \frac{\pi^2}{h^2}\left(9\,b^2_{mk} + \frac{9}{2}\,b^2{}_{lk}\right).\frac{\tau_c}{1+4.\pi^2\nu^2\tau_c{}^2}$$

$$\left.\vphantom{\begin{array}{c}1\\1\\1\end{array}}\right\}\ \ldots(4.27)$$

b_{lk} and b_{mk} are constants derived from the time average over the wave functions and have orders of magnitude of $g.\beta.g_N.\beta_N.\left[\dfrac{1}{r_k{}^3}\right]_{\mathrm{Av.}}$, where r_k is the distance between the unpaired electron and the proton. The parameter τ_c is 'the correlation time' for the angular functions $(\cos^2\theta - \frac{1}{3})$ and $\sin\theta\cos\theta$. At microwave frequencies the factor $\tau_c/(1+4\pi^2\nu^2\tau_c{}^2)$ becomes small and the line width is then determined mainly by T_2', and typical values of τ_c are found to be 10^{-8} sec. The value of τ_c can also be derived from the Debye theory, as in the treatment of Bloembergen, Purcell and Pound[27]. It can be shown from this that

$$\tau_c = 4\pi.\eta.a^3/3kT \qquad \ldots(4.28)$$

where η is the viscosity of the medium and a is the effective radius of the free radical. It can be seen that if the viscosity of the medium increases so does the value of the 'correlation time', τ_c. It follows from these quantitative expressions that it is possible, in principle, to deduce the viscosity and nature of the medium in which the radicals are held from the nuclear dipolar contribution to the line width.

Another way in which the dipolar broadening may be partially reduced is when the molecules of the specimen under investigation are held in a rigid matrix, while internal motion of individual groups of the molecule, such as $-CH_3$ rotation, still occurs. In this case the protons, or other magnetic nuclei, of the neighbouring molecules will still contribute their full dipolar broadening, while that from the moving protons within the same molecule will be averaged out. This mechanism probably accounts for the reduction of line width from 25 to 10 gauss for most radicals trapped in low temperature hydrocarbon glasses[16].

4.4.2 Spin-lattice Interaction

Spin-lattice interaction is the name given to all the processes in which the electron spins can share energy with the molecule, or lattice, as a whole. Magnetic energy is thus converted into thermal energy, and a relaxation process is produced by which electron spins in the excited level can return to the ground state. The most general form of spin-lattice interaction takes place via the spin-orbit coupling since the thermal vibrations of the ' lattice ', or molecule, will only couple directly to the orbital motion of the electron, in the first instance. The theory of spin-lattice interaction acting via the effect of a modulating electrostatic field has been considered in detail by KRONIG[28]. He postulated two mechanisms whereby the spins could exchange energy with the lattice, either by exchanging a whole quantum directly with a lattice vibration of the appropriate frequency, or by scattering a quantum from the lattice, and changing its value. Both of these mechanisms predict a spin-lattice relaxation time summarized in equation 4.29, where Δ is the splitting to the next orbital level, λ is the spin-orbit coupling coefficient, and T is the absolute temperature. M has a value of 4 or 6, and N a value of 1 or 7.

$$T_1 \propto \frac{\Delta^M}{\lambda^2 T^N} \qquad \dots (4.29)$$

In free-radicals the spin-orbit coupling coefficient is small, and due to the strong quenching of orbital motion by the molecular asymmetry, the value of Δ is very large, and as a result very long values of spin-lattice relaxation time can be obtained. Thus, in normal paramagnetic atoms, with relatively high spin-orbit interaction, T_1 is usually of the order of 10^{-8} sec, whereas in free-radicals it may be as long as several seconds. The spin-lattice interaction via the orbital coupling is thus very weak in most free radical studies[29, 30] and does not produce a significant broadening of the absorption lines.

126

In certain cases, however, other mechanisms may be effective in sharing energy between the spin system and the lattice. Thus the tumbling motion which affects the dipolar spin-spin interaction will also allow transfer of magnetic to thermal energy[26]. Any exchange interaction that is present will also allow transfer of energy from the spin system to the exchange energy which is, in turn, closely coupled to the lattice[31]. In such cases the main contribution to the spin-lattice interaction will be from these processes, instead of the spin-orbit coupling.

It is probably true to say, however, that the spin-lattice interaction in free radicals is never strong enough to cause appreciable broadening by shortening the lifetime of the excited state. The reverse is in fact often true, i.e. that the interaction is not strong enough to allow sufficiently rapid thermal relaxation, and ' saturation broadening ' therefore results.

4.4.3 *Exchange Narrowing*

Exchange interaction has been discussed qualitatively in the previous sections and it has been seen that this occurs when electrons are exchanged between the orbitals of different molecules, and is a direct consequence of the Uncertainty Principle. It is possible for this exchange, or ' spin-dependent coupling ' to take place even when there is negligible chemical binding between the molecules, and it is a very common feature of free radical spectra whenever high concentrations of radicals are present.

The theory of exchange interaction has been treated in detail by Van Vleck[24], Pryce and Stevens[25], and ANDERSON and WEISS[32] and KIVELSON[37]. They have shown that if the exchange is between similar ions or molecules it will narrow the absorption line in the centre and broaden it in the wings, leaving the second moment $(\Delta H)^2_{Av.}$ unchanged, but reducing the width measured at half-power points. It may be noted in this connection that for a Gaussian line shape, as is obtained from spin-spin interaction alone, the line width as measured between the points of maximum slope is equal to 0·852 times the width measured at half-height, and that both of these are reduced by exchange interaction.

The theory of Anderson and Weiss[32] attempted to derive a detailed line shape in the case of strong exchange by using a model of ' random frequency modulation '. This predicted a line of ' resonance ' or Lorentzian shape in the centre, falling off more rapidly in the wings, and appears to be in good agreement with the experimental results. The change of an absorption line from a Gaussian to Lorentzian shape, can be taken as very good evidence that exchange narrowing has taken place. The equations for both these line shapes are given below, for reference purposes, and are illustrated in *Figure 35(a)* and *(b)*.

<table>
<tr><td align="center">*Gaussian Line*</td><td align="center">*Lorentzian Line*</td></tr>
</table>

Normalized Equation

$$g(\omega-\omega_0)=\frac{T_2}{\pi}\cdot e^{-(\omega-\omega_0)^2\cdot T^2/\pi} \qquad g(\omega-\omega_0)=\frac{T_2/\pi}{1+(\omega-\omega_0)^2 T_2^2}$$

$$\dots\dots(4.30 \text{ and } 4.31)$$

Width at half-height

$$(\pi.\log_e 2)^{1/2}\cdot\frac{1}{T_2} \qquad\qquad \frac{1}{T_2} \qquad \dots\dots(4.32 \text{ and } 4.33)$$

Width at point of maximum slope

$$\frac{1}{T_2}\sqrt{\frac{\pi}{2}} \qquad\qquad \frac{1}{T_2\sqrt{3}} \qquad \dots\dots(4.34 \text{ and } 4.35)$$

Note. In these equations T_2 is defined as the inverse of the line width parameter and is equal to π times the maximum value of $g\,(\omega-\omega_0)$.

Another line narrowing process may be mentioned here, since although it is not an exchange interaction, they have several features in common. This is the motional narrowing due to the delocalization of the unpaired electrons in a molecular orbital. In free radicals the unpaired electron will pass over several atoms in its orbit, and this motion may produce an averaging of the dipolar interactions and a line narrowing similar to that of exchange or liquid motional narrowing. This possibility has been investigated in some detail by WALTER *et al*[33]. who concluded that there is definite experimental evidence for this ' delocalization narrowing '.

4.5 SATURATION BROADENING

It has been seen in the preceding sections that free radicals are often characterized by long spin-lattice relaxation times. It was also noted in the introductory section 1.6 that such long relaxation times can produce a ' saturation broadening ' because the spins in the excited level cannot return to the ground level sufficiently quickly. This effect will now be considered quantitatively.

4.5.1 *Homogeneous Broadening*

The two electron levels are denoted by A and B, where A is the excited level, and the numbers in the two levels are denoted by N_1

and N_2 per gramme respectively. In thermal equilibrium

$$\frac{N_1}{N_2} = e^{-\hbar.\omega_0/k.T_L} \qquad \dots (4.36)$$

where T_L is the temperature of the lattice.

Let $n = N_2 - N_1$, and when microwave resonance radiation is applied

$$\frac{dn}{dt} = \left(\frac{dn}{dt}\right)_{r.f.} + \left(\frac{dn}{dt}\right)_{s.l.} \qquad \dots (4.37)$$

The first term on the right is the rate of transitions produced by the microwave field and is given by radiation theory as

$$\left(\frac{dn}{dt}\right)_{r.f.} = -\frac{1}{4}\pi.\gamma^2.H_1^2.g(\omega-\omega_0).n. \qquad \dots (4.38)$$

where γ is the gyromagnetic ratio of the electron, H_1 is the strength of the microwave magnetic field and $g(\omega-\omega_0)$ is the shape function of the absorption line. This is normalized so that

$$\int_0^\infty g(\omega-\omega_0).d\omega = 1 \qquad \dots (4.39)$$

and is represented in *Figure 35(c)*.

The second term on the right in equation 4.37 is the rate of transitions produced by the spin-lattice interaction, and this is given by

$$\left(\frac{dn}{dt}\right)_{s.l.} = \frac{n_0-n}{T_1} \qquad \dots (4.40)$$

where n_0 is the value of n at thermal equilibrium and is given from equation 4.36 as

$$n_0 = N_{2_0}-N_{1_0} \approx \frac{\hbar.\omega_0}{kT_L}.N_{1_0} \qquad \dots (4.41)$$

When equilibrium conditions have set in $\left(\dfrac{dn}{dt}\right)_{\text{total}}$ must be zero, and hence

$$-\frac{1}{4}\pi.\gamma^2.H_1^2.g(\omega-\omega_0).n+\frac{n_0-n}{T_1} = 0 \qquad \dots (4.42)$$

$$\therefore n = n_0(1+\frac{1}{4}\pi.\gamma^2.H_1^2.g(\omega-\omega_0).T_1)^{-1} \qquad \dots (4.43)$$

The rate at which energy is absorbed from the magnetic field may now be calculated.

Thus
$$P_a = -\hbar \cdot \omega \cdot \left(\frac{dn}{dt}\right)_{r.f.} \qquad \dots (4.44)$$

and substituting 4.38 and then 4.43 into this gives

$$P_a = \hbar \cdot \omega \cdot \frac{1}{4} \cdot \pi \cdot \gamma^2 \cdot H_1^2 \cdot g(\omega - \omega_0) \cdot n_0 \cdot \left[1 + \frac{1}{4}\pi \cdot \gamma^2 \cdot H_1^2 \cdot g(\omega - \omega_0) T_1 \right]^{-1}$$

$$= \tfrac{1}{2}\omega \cdot \omega_0 \cdot \left(\frac{\gamma^2 \cdot \hbar \cdot n_0}{2\,\omega_0}\right) \cdot H_1^2 \cdot \left[\frac{\pi \cdot g(\omega - \omega_0)}{1 + \frac{1}{4}\pi \cdot \gamma^2 \cdot H_1^2 \cdot T_1 \cdot g(\omega - \omega_0)}\right]$$

$$\dots (4.45)$$

This rate of energy absorption is also given by the expression for the complex susceptibility, see equation 2.8.

$$P_a = \frac{1}{2}\,\omega \cdot \chi'' \cdot H_1^2 \qquad \dots (4.46)$$

Therefore, equating 4.45 and 4.46, and substituting χ_0 for $\dfrac{\gamma^2 \cdot \hbar \cdot n_0}{2\,\omega_0}$

$$\chi'' = \omega_0 \cdot \chi_0 \cdot \left[\frac{\pi \cdot g(\omega - \omega_0)}{1 + \frac{1}{4}\pi \cdot \gamma \cdot {}^2H_1^2 \cdot T_1 \cdot g(\omega - \omega_0)}\right] \qquad \dots (4.47)$$

At the resonance frequency, $g(\omega - \omega_0)$ will have its maximum value, and from the definition of T_2 in equation 4.25 this maximum value is equal to $T_{2/\pi}$. Hence

$$\chi''_{\omega_0} = \chi_0 \cdot \omega_0 \cdot T_2 \cdot \frac{1}{(1 + \frac{1}{4}\gamma^2 \cdot H_1^2 \cdot T_1 \cdot T_2)} \qquad \dots (4.48)$$

It will be seen by comparison with equation 2.17a that, since $T_2 = \dfrac{1}{\varDelta\omega}$, in the absence of saturation, χ''_{ω_0} is given by

$$\chi''_{\omega_0} = \chi_0 \cdot \omega_0 \cdot T_2 \qquad \dots (4.49)$$

and the term ' saturation factor, Z ', is therefore given to the remaining expression of 4.48.

I.e.
$$Z = \frac{n_s}{n_o} = \frac{1}{\left(1 + \frac{1}{4} \cdot \gamma^2 \cdot H_1^2 \cdot T_1 \cdot T_2\right)} \qquad \dots (4.50)$$

This equation implies that the absorption of power will be lower the longer the value of T_1, and that the fall-off in expected power absorption will increase with increasing levels of input power (H_1). Moreover this decrease in expected absorption will occur first in the centre of the lines where the greatest power is absorbed, and only affect the wings as the value of H_1 rises still further. Hence this saturation effect, produced by large values of T_1 and H_1, will not only reduce the actual power absorption but also alter the line shape, flattening it in the centre before the wings, and thus increasing the apparent width. This fact may be seen quantitatively in equation 4.47, since $g(\omega - \omega_0)$ is multiplied by H_1 in the denominator, but not in the numerator, and it therefore follows that the line will change shape as H_1 increases. The detailed estimation of saturation broadening in any particular case is best accomplished by drawing out the curve of equation 4.47 for different line shapes and values of H_1, and then comparing these with those obtained experimentally.

The theory outlined above applies to all cases of ' homogenous broadening '—i.e. cases where the line is initially broadened by interactions within the spin system, or from an external interaction which is fluctuating rapidly compared with the time taken for a spin transition[34]. It therefore includes dipolar spin-spin interaction; spin-lattice interaction, and motional or exchange narrowing.

4.5.2 *Inhomogeneous Broadening*

Although the theory of saturation given above applies to most interactions which broaden free-radical resonance lines, there is a class of interaction for which some modifications of the treatment is necessary. These interactions are those which come from outside the spin system and which vary slowly in time compared with a spin transition. Such ' inhomogeneous broadening ' is produced by unresolved hyperfine components; and inhomogeneities of the magnetic field. It is evident that in both of these cases the total line shape is the envelope of several real absorption lines. Saturation will therefore occur separately for these component lines as outlined above, and as a result each will have its peak height reduced by the same saturation factor. The envelope of the lines will therefore retain the same shape, and hence the ' inhomogeneously broadened ' line will not change shape or width on saturation, but the expected power absorption will fall in the same proportion across its whole width.

The theory of inhomogeneous broadening has been treated in detail by PORTIS[34], and the expression he obtained for $\chi''(\omega)$ does not contain ω in the integrated shape parameter, confirming that the line shape does not change on saturation. It follows that a careful study of change in absorbed power, and any change in width, on saturation, allows the mechanism of broadening to be determined, and the presence of any unresolved hyperfine structure to be detected.

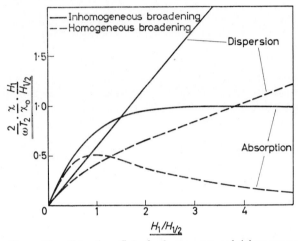

Figure 36. Saturation effects for homogeneous and inhomogeneous broadening. Observed susceptibility plotted against microwave field strength (After Portis[34])

The quantitative difference in power absorption for the two cases is summarized in *Figure 36*, which is taken from the paper by Portis[34]. The parameter plotted on the ordinate is equal to $\dfrac{2}{\omega T_2} \cdot \dfrac{\chi}{\chi_0} \cdot \dfrac{H_1}{H_{1/2}}$ and that along the abcsissa is $H_1/H_{1/2}$ where $H_{1/2}$ is the value of the microwave field strength which makes the saturation parameter equal to one half.

4.5.3 *Experimental determination*

It has been seen above that the change of line shape on saturation can be used to differentiate between different types of broadening. It is also possible to calculate the value of the spin-lattice relaxation time from the decrease in absorbed power. The ratio of actual power absorbed to that expected with the given microwave magnetic field is equal to the saturation factor, Z, of equation 4.50. This ratio can be determined simply by plotting the magnitude of the absorbed power against H_1^2 and noting its deviation from a linear plot. It therefore follows that, since γ and T_2 are known, T_1 can be calculated from the values of Z and H_1.

132

The value of the microwave field at the specimen, H_1, can be calculated if the Q of the cavity and the power entering the cavity from the waveguide are known. Thus it may be shown that the total power down a waveguide carrying an H_{01} mode is given by

$$P_g = \frac{30\pi}{\sqrt{\left[1-\left(\frac{\lambda}{2a}\right)^2\right]}} \cdot ab. \; H^2_{max.} \text{ watts} \qquad \ldots(4.51)$$

Therefore, if the cavity has a factor Q, the value of the peak microwave magnetic field inside can be written as

$$H^2_{max.} = Q.P_g. \frac{\sqrt{1-\left(\frac{\lambda}{2a}\right)^2}}{30\pi. \; a.b.} \text{ gauss}^2 \qquad \ldots(4.52)$$

where P_g is the power in the waveguide in watts, and the guide dimensions a and b, are in centimetres. For standard X-band waveguide (0·9 inches by 0·4 inches), and a wavelength of 3·2 cm, this equation becomes

$$H_{max.} = \frac{1}{17\cdot5}\sqrt{Q.P_g.} \text{ gauss} \qquad \ldots(4.53)$$

The Q factor of the cavity can be determined either directly from the spacing of frequency markers on the oscilloscope screen, or by the measurement of standing wave ratios over a small frequency band[35]. A method very suitable for reflection type cavities has recently been described by SMIDT[36], in which readings of the crystal current from the fourth arm of the bridge, at different frequency settings, are the only measurements that have to be taken.

REFERENCES

[1] ABRAGAM, A. and PRYCE, M. H. L. *Proc. roy. Soc. A* 205 (1951) 135
[2] WEISSMAN, S. I. and BANFILL, D. *J. Amer. chem. Soc.* 75 (1953) 2534
[3] CHU, T. L., PAKE, G. E., PAUL, D. E., TOWNSEND, J. and WEISSMAN, S. I. *J. phys. chem.* 57 (1953) 504
[4] JARRETT, H. S. and SLOAN, G. J. *J. chem. Phys.* 22 (1954) 1783
[5] TUTTLE, T. R., WARD, R. L. and WEISSMAN, S. I. *Ibid.* 25 (1956) 189
[6] BLEANEY, B. and INGRAM, D. J. E. *Proc. roy. Soc. A.* 205 (1951) 336
[7] WEISSMAN, S. I. *J. chem. Phys.* 25 (1956) 890
[8] McCONNELL, H. M. *Ibid.* 24 (1956) 764
[9] JARRETT, H. S. *Ibid.* 25 (1956) 1289
[10] BERSOHN, R. *Ibid.* 24 (1956) 1066
[11] TUTTLE, T. R. and WEISSMAN, S. I. *Ibid.* 25 (1956) 189
[12] ALTMANN, S. L. *Proc. roy. Soc. A,* 210 (1951) 327, 343

REFERENCES

[13] Sogo, P. B., Nakazaki, M. and Calvin, M. *J. chem, Phys.* 26 (1957) 1343
[14] McConnell, H. M. and Chestnut, D. B. *Ibid.* 27 (1957) 984; 28 (1958) 107
[15] Lefkovitz, H. C., Fain, J. and Matsen, F. A. *Ibid.* 23 (1955) 1690
[16] Fujimoto, M. and Ingram, D. J. E. *Trans. Faraday Soc.* 54 (1958) 1304
[17] Coulson, C. A. *Valence.* University Press, Oxford, 1953, p. 312
[18] Andrew, E. R. *Nuclear Magnetic Resonance.* University Press, Cambridge, 1955, p. 155
[19] Andrew, E. R. and Bersohn, R. *J. chem. Phys.* 18 (1950) 159
[20] Waugh, J. S., Humphrey, F. B. and Yost, D. M. *J. phys. Chem.* 57 (1953) 486
Bersohn, R. and Gutowsky, H. S. *J. chem. Phys.* 22 (1954) 651
[21] Gordy, W., Ard, W. B. and Shields, H. *Proc. nat. Acad. Sci., Wash.* 41 (1955) 983
[22] Pake, G. E. *J. chem. Phys.* 16 (1948) 327
[23] Bleaney, B., Elliott, R. J. and Scovil, H. E. D. *Proc. phys. Soc. A,* 64 (1951) 933
[24] Van Vleck, J. H. *Phys. Rev.* 74 (1948) 1168
[25] Pryce, M. H. L. and Stevens, K. W. H. *Proc. phys. Soc. A* 63 (1950) 36
[26] McConnell, H. M. *J. chem. Phys.* 25 (1956) 709
[27] Bloembergen, N., Purcell, E. M. and Pound, R. V. *Phys. Rev.* 73 (1948) 679
[28] Kronig, R. de L. *Physica,* 6 (1939) 33
[29] McClure, D. S. *J. chem. Phys.* 20 (1952) 682
[30] Mizushima, M. and Koide, S. *Ibid.* 20 (1952) 765
[31] Bloembergen, N. and Wang, S. *Phys. Rev.* 93 (1954) 72
[32] Anderson, P. W. and Weiss, P. R. *Rev. mod. Phys.* 25 (1953) 269
[33] Walter, R. I., Codrington, R. S., D'Adamo, A. F. and Torrey, H. C. *J. chem. Phys.* 25 (1956) 319
[34] Portis, A. M. *Phys. Rev.* 91 (1953) 1071
[35] Montgomery, C. G. *Technique of Microwave Measurements.* McGraw-Hill, 1947, p. 333
King, D. D. *Measurements at Cm Wavelengths.* Van Nostrand, 1952, p. 128
[36] Smidt, J. *Appl. sci. Res. Hague, B* 6 (1957) 353
[37] Kivelson, D. *J. chem. Phys.* 27 (1957) 1087.

134

STABLE FREE RADICALS

5.1 INTRODUCTION

MOST interest in free radical studies is probably centred in short-lived active species and the kinetic processes associated with them. There are, however, quite a large number of stable, or long-lived, radicals and a study of these is very useful as a preliminary to an understanding of the more active types. Stable radicals can usually be studied over a wide range of concentration, and very detailed hyperfine patterns are often obtained from dilute solutions. The detailed molecular orbital associated with the unpaired electron can often be derived and the general theories of hyperfine interaction then tested. In this way the results obtained on stable radicals have served as a very useful proving ground for various theories. These theories can then be applied to other radical studies to give much more information than would otherwise have been possible.

The results obtained on stable free radicals are therefore considered first in this book, and are summarized in this chapter. It is shown how each set of results has been used to confirm or extend the theoretical interpretation of the free radical interactions, and this background of experience is then used when analysing the results of the more active radicals discussed in the following chapters. The study of stable radicals has some very interesting features in its own right, as well, and these are considered for each particular system separately. The stable organic radicals are divided into five groups in this chapter, the classification being more by the type of electron resonance spectra obtained than by their chemical constitution, although in many respects the two go together. The five groups are as follows: (*i*) Diphenylpicrylhydrazyl and its related derivatives; (*ii*) Aromatic ions in solution; (*iii*) Semi-quinones; (*iv*) Triphenylmethyl, Dimesitylmethyl and Perinaphthene; and (*v*) Radicals containing sulphur. Hydrazyl and its derivatives are treated first since many more measurements have been made on this radical than any other, and it serves as a very good illustration of the theories of g-value variation, line width interactions, and hyperfine splitting. The aromatic negative ions are considered next as the very highly resolved hyperfine patterns obtained from these have allowed the theories of configurational interaction to be checked in detail quantitatively. The semiquinones have also given some very interesting hyperfine patterns and served to confirm the theory of hyperfine interaction by hyperconjugation. The triphenylmethyl, dimesitylmethyl and perinaphthene radicals are classed together and treated next since they appear at first sight to contradict the generalized theory of configurational interaction as applied to aromatic radicals, and the concept of 'negative spin density' has to be used to account for their spectra. The rather different type of spectra obtained from radicals containing sulphur are then considered at the end, together with a variety of other species.

It may be noted that stable *inorganic* radicals such as those in metal-ammonia solutions, or those trapped at low temperatures from discharge tube reactions, are considered separately in Chapter 8.

5.2 DIPHENYLPICRYL HYDRAZYL AND SIMILAR DERIVATIVES

The stable crystalline radical 1, 1-diphenyl-2-picryl hydrazyl was one of the first to be studied by electron resonance[1], and has been used as a reference signal in most electron resonance spectrometers ever since. Its preparation is a relatively simple matter[2] and crystals up to $\frac{1}{2}$ cm long can be obtained. It has the formula

The electron resonance signal obtained from the polycrystalline form has a width at half height of 2·7 gauss and since each molecule has one unpaired electron associated with it, a very intense signal is obtained. It is in fact possible to detect less than 10^{-9} grammes in a room temperature X-band spectrometer[3]. The resonance line has several features of interest, such as a slightly anisotropic g-value and line width, strong exchange narrowing, and a hyperfine structure from the nitrogen atoms when in dilute solution. These will be considered separately as follows.

5.2.1 *g-value*

The mean g-value obtained from a polycrystalline sample is $2\cdot0036\pm0\cdot0003$ and this is often used as a standard reference for the determination of unknown g-values close to 2·0. The value is significantly different from the free-electron spin value of 2·0023, showing that there is a small residual spin-orbit coupling present. Of considerable interest, however, is the anisotropy of the g-value and its temperature dependance. Initial measurements by KIKUCHI and COHEN[4,5] showed that if a single crystal were rotated about an axis normal to the broad face a $(3 \cos^2 \theta - 1)$ variation of the g-value was obtained. The magnitude of this variation was found to depend on the strength of the applied d.c. magnetic field, and it was therefore suggested that the anisotropy was connected with the diamagnetism of the benzene rings. Further measurements[6] were then made at various temperatures down to $1.6°$ K, and the anisotropy was found to increase markedly at the lower temperatures. Thus at room temperature the extreme g-values were measured as 2.0028 and 2.0038; at $77°$ K as 2.0027 and 2.0039; at $4.2°$ K as 2.0037 and 2.0067; and at $1.6°$ K as 2.005 and 2.01. It is thought[14] that this anisotropy and its increase at low temperatures is due to the effect of the finite susceptibility of the benzene ring system, and the reduction of the motional exchange narrowing of the delocalized orbitals.

5.2.2 Line Width

The unpaired electrons in crystalline hydrazyl are quite close together, with an average separation of about $10Å$, and hence the dipolar spin-spin interaction between them should produce a broadening of the resonance line of about 30 gauss. The fact that the measured width is only 2.7 gauss indicates that strong exchange narrowing must be present, and the Lorentzian, rather than Gaussian, shape of the absorption line confirms this. It is in fact very reasonable to expect such an exchange interaction when the molecular orbitals containing the unpaired electrons are so close. A direct measurement of the mechanism responsible for the line width of the hydrazyl resonance has been made by BLOEMBERGEN and WANG[7], employing a 'saturation' technique. By the use of pulsed magnetron power sources they were able to saturate the resonance and thus measure T_1 from the saturation parameter. It was necessary to produce microwave magnetic field strengths of 2 gauss at the sample before 50 per cent saturation occurred, and T_1 was determined as $6.3 . 10^{-8}$ sec. This spin-lattice relaxation time was found to be independent of temperature and the spin-lattice interaction was therefore postulated as acting directly via the exchange energy, and not via the spin-orbit coupling, in this case. LLOYD and PAKE[8] also made saturation measurements on hydrazyl at radiofrequencies where it was easier to produce high intensity values of H_1, and obtained a very similar value for the spin-lattice relaxation time. LIVINGSTONE and ZELDES[9] have also reported that there is a small anisotropy in line width of single crystals of hydrazyl as well as in the g-value.

A considerable amount of work has also been carried out on both crystalline hydrazyl and its solutions at low radiofrequencies. The detailed theory of line shape at very low field strengths has been given by GARSTENS[10] and his predictions have been checked by several experiments[10,11].

5.2.3 *Hyperfine Interaction*

If the hydrazyl is dissolved in benzene, the line width initially broadens out as the exchange narrowing is reduced, and, on further dilution, the electron spin-spin interaction is also decreased and a resolved hyperfine pattern is obtained from the nuclear interaction. This hyperfine pattern is shown in *Figure 37(a)*, as it appears on the oscilloscope screen and five lines with a symmetrical distribution in intensity can be clearly seen. This structure was first obtained by HUTCHISON, PASTOR and KOWALSKY[12] in a 0·002 molar solution in benzene. They interpreted the five line spectrum as arising from an equal coupling of the unpaired electron to the nuclei of the two central nitrogen atoms of the hydrazyl. The nitrogen nuclei have a nuclear spin of $I=1$ and hence each electronic level will be split into $(2I+1)=3$ components by an interaction with one nitrogen nucleus. Such a splitting is shown in the central column of *Figure 37(c)* which represents the energy level system for the hydrazyl radical. If the interaction with the other nitrogen were much smaller than with the first then each of these three levels would be split into another three sub-components of smaller spacing, and there would therefore be a total of nine components in each electronic level. The fact that only five instead of nine lines are observed in the hydrazyl hyperfine pattern suggests that there must be some overlap of these components. If the two nitrogen atoms are coupled equally to the electron then the interaction with the second nitrogen will split the initial components into a further triplet, with a splitting equal to that produced by the first nitrogen atom. The resulting energy level system will therefore be as shown in *Figure 37(c)* and it will be seen that the three central components are formed of overlapping levels. The relative strength of the five components, as measured by the number of levels included in them, will be $1 : 2 : 3 : 2 : 1$, and this will also be the relative intensities of the resulting absorption lines. An equal interaction between the unpaired electron of the free radical and the two nitrogen atoms will therefore produce a five-line spectrum, as observed for the hydrazyl,

138

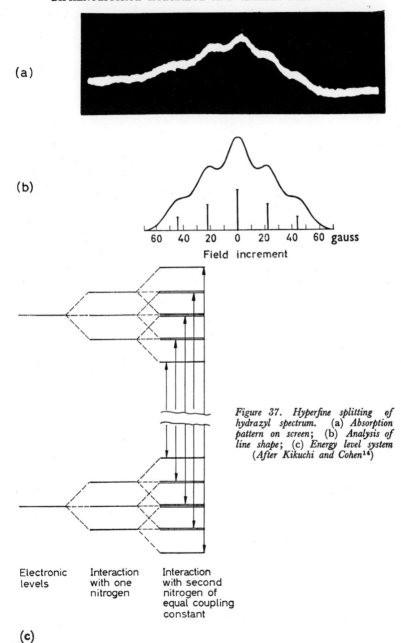

(a)

(b)

Field increment

Figure 37. Hyperfine splitting of hydrazyl spectrum. (a) Absorption pattern on screen; (b) Analysis of line shape; (c) Energy level system (After Kikuchi and Cohen[14])

Electronic levels   Interaction with one nitrogen   Interaction with second nitrogen of equal coupling constant

(c)

139

Structural formula of carbazyl

(a)

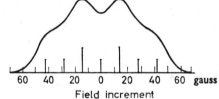

(b)

(i)
Observed
spectrum

(ii)
Analysis
of
line shape

60 40 20 0 20 40 60 **gauss**

Field increment

(c)

Electronic   Interaction   Interaction with
level   with one   second nitrogen
                    nitrogen   of half coupling
                                         constant

140

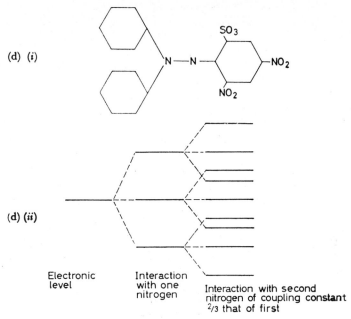

Figure 38. *Hyperfine splitting of carbazyl.* (a) *Structural formula*; (b) *Absorption pattern, and analysis of line shape*; (c) *Energy level system*; (d) (i) *Structural formula of diphenyldinitrosulphophenyl hydrazyl*; (ii) *Energy level system (After Jarrett[13])*

and the relative intensitities are also in agreement with those predicted.

This hyperfine pattern may be compared with that obtained from two other radicals, very similar in structure to 1,1-diphenyl-2-picryl hydrazyl, but giving strikingly different spectra. One of these is carbazyl (N-picryl-9-aminocarbazyl) which has a bridge between the two phenyl groups but is otherwise similar to the 1,1-diphenyl-2-picryl hydrazyl. The other is the sodium salt of diphenyl-dinitro-sulphophenyl hydrazyl, in which one of the NO_2 groups of the picryl ring has been replaced by an SO_3 group. The structural formula for carbazyl is given in *Figure 38(a)*, with the observed hyperfine pattern and its analysis in *Figure 38(b)*. The energy level splitting consistent with this is shown in *Figure 38(c)*, only the splitting of one electronic level is included in this case, since the other will be identical but inverted. These spectra were first obtained by JARRETT[13] and also investigated by KIKUCHI and COHEN[14]. It is seen that the hyperfine spectrum from the carbazyl consists of seven lines with an intensity distribution of approximately $1:1:2:1:$

2 : 1 : 1. This pattern may be explained if it is assumed that the interaction with one nitrogen atom is exactly twice that with the other. The splitting produced by the stronger interaction is drawn first in the centre of *Figure 38(c)* followed by that produced by the weaker. The triplet of the central level will thus have both of its outer components overlapping with those from the levels above and below, and the seven sets of levels are thus explained, with the pair on either side of the centre having a strength of two.

The full total of nine hyperfine lines are observed for the sulphon-ated derivative and this implies that the interactions with the two nitrogens have no simple integral relationship, and no overlapping of the component levels thus occurs. The pattern can be explained in detail by the energy level system shown in *Figure 38(d)*, and it may be noted that the two different splittings correspond to 12 gauss (33 Mc/s) and 8 gauss (23 Mc/s). The quantitative distribution of the molecular orbital of the unpaired electron can thus be derived directly from the measured splittings of the hyperfine pattern.

These results are a good example of the marked change in spectrum that small changes in radical structure may produce, and show how sensitive a probe the hyperfine splitting can be into the detailed shape of the electron's molecular orbit. No hyperfine components arising from an interaction with the edge protons are observed in this case, suggesting that the unpaired electron is con-centrated near the centre of the radicals. On the other hand BELJERS, VAN DER KINT and VAN WIERINGEN[15] have detected an Overhauser effect for these protons, showing that there must be some small coupling to the unpaired electron.

The case of the hyperfine pattern of hydrazyl and its related derivatives serves as a good introduction to the more complex spectra. Thus in the hydrazyls the spectra is produced by an interaction with only two nuclei whereas in the aromatic systems to be considered below, the interaction is often with a large number of nuclei, and the hyperfine patterns tend to become very involved.

5.3 AROMATIC IONS IN SOLUTION

5.3.1 *Negative Ions*

A large number of stable free radicals can be formed as ions in solution by the reaction of metallic sodium or potassium with aromatic compounds. The reaction takes place more rapidly in good electron-transfer solvents such as dimethoxyethane or tetra-hydrofuran, and can be represented for a compound such as naphthalene by

$$C_{10}H_8 + Na \longrightarrow C_{10}H_8^- + Na^+ \qquad \qquad \dots (5.1)$$

The naphthalene negative ion can therefore be considered as a naphthalene molecule with an extra unpaired electron added, which moves in a molecular orbital over the rings. Such aromatic negative ions have been studied in considerable detail by electron resonance and have provided a very useful field for testing quantitative theories of hyperfine interaction. The case of the naphthalene negative ion will be considered in detail first, as an example of this, and the results obtained on other aromatic compounds will then be summarised more briefly.

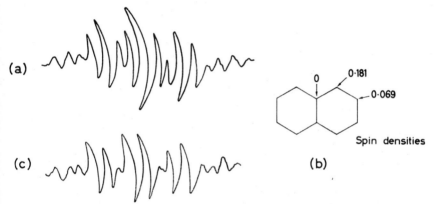

Figure 39. Hyperfine pattern of the naphthalene negative ion (derivative tracing). (a) Negative ion in dimethoxyethane; (b) Calculated spin densities; (c) Spectrum of β-deuterated ion (After Weissman et al.[16–19])

The hyperfine splitting of the naphthalene negative ion was first studied by WEISSMAN, TOWNSEND, PAUL and PAKE[16–18] who employed 10^{-3} molar solutions in tetrahydrofuran, and observed seventeen hyperfine components with a successive separation of about 1·5 gauss. A further study of this[19] showed that exactly the same spectrum, as illustrated in Figure 39(a) could be obtained from the salts of lithium or potassium in tetrahydrofuran, and of sodium, potassium and rubidium in dimethoxyethane. The simple molecular orbital treatment of HÜCKEL[20] can be applied to the naphthalene molecule to calculate the expected spin densities on the different carbon atoms, and the result of this is shown in Figure 39(b).

It is seen that there are only two different kinds of proton position in the naphthalene molecule, and hence the hyperfine pattern should be the combination of an interaction with four equivalent α-hydrogens, giving a five-line spectrum, and an interaction with four β-hydrogens, which will split each of the initial lines into another five components. The general case of a non-integral relation between the coupling constants of the two protons, would therefore produce twenty-five hyperfine components of equal intensity. It can be seen from the figures derived from the simple Hückel model[20], however, that the ratio of the spin densities on the two types of proton is nearly 3·0. If an energy level system is now constructed, in a way similar to that used for the hydrazyls in the last section, but for two successive five fold splittings with a ratio of 3 : 1, it will be seen that a seventeen line pattern is obtained with relative strengths of 1 : 1 : 1 : 2 : 2 : 1 : 2 : 2 : 1 : 2 : 2 : 1 : 2 : 2 : 1 : 1 : 1. This predicted pattern agrees very well with the observed spectrum, and the three singlet components can be clearly seen at each end. In fact the splitting ratio is not exactly three, and the small deviations in the spectrum from seventeen equally-spaced lines can be explained if the ratio is taken as 2·8. The values of the two splitting constants are then 1·79 and 5·01 gauss. It is therefore seen that the detailed analysis of the hyperfine pattern serves as a very precise check on the molecular orbital calculations. It may be noted that the overall splitting is 27·2 gauss in good agreement with the 28 gauss predicted by the configurational interaction treatment of the last chapter.

A further check on this interpretation of the spectrum and an assignment of the two splitting constants between the α and β protons can be made by studying the deuteronaphthalene negative ions[19,21]. The deuterium was substituted either on to one of the α, or on to one of the β positions, and the different hyperfine patterns were recorded. A spectrum of fifteen lines is to be expected if the deuterium is in a high coupling-constant position, whereas a spectrum of sixteen lines is expected if it is in the low coupling-constant position. The expected intensity distributions between the lines will be very different and there is no ambiguity in distinguishing the two spectra. When the experiments were performed it was found that these predicted fifteen and sixteen line spectra were obtained, the former for the α and the latter for the β deutero compounds. The spectrum for the β deuteronaphthalene ion is shown in *Figure 39(c)*.

It was shown in the last chapter that the configurational interaction would not only allow the π orbital of the unpaired electron to admix with some of the s state wave function of the adjoining protons, and thus produce isotropic hyperfine structure, but also predicts a finite density for it at the nuclei of the aromatic ring itself. This normally has no effect on the hyperfine pattern since C^{12} has no nuclear spin nor magnetic moment. If C^{13} is substituted for the C^{12}, however, an extra splitting due to the interaction with its spin of $\frac{1}{2}$ is to be expected, and the observation of this was one of the striking confirmations of the theory of configurational interaction. TUTTLE and WEISSMAN[22] studied a naphthalene negative ion in which 53 per cent of the C^{12} in the α position had been replaced by C^{13}. The spin of $\frac{1}{2}$ of the C^{13} should split each of the normal hyperfine components into two, thus producing 34 components. The actual observed spectrum was a superposition of this on the original 17 line spectrum, the ratio of the intensities of the two corresponding to the C^{13} to C^{12} ratio. By a comparison of the predicted and observed patterns the splitting due to the C^{13} was deduced as 7·1 gauss, from which the spin density on the carbon nucleus can be calculated. This technique of C^{13} substitution is a very powerful tool in analysis of free radical structure, and other examples will be found in the following sections.

The implications of the electron resonance work on the naphthalene negative ion can therefore be summarized as follows.

(i) It has proved conclusively that the action of sodium on the aromatic compounds is initially a one-electron and not a two-electron process.

(ii) The molecular orbital calculations for the system can be checked quantitatively from the hyperfine pattern, and more precise numerical factors obtained from the splitting parameters (e.g. Hückel's theory gives a ratio of 2·6 for the α and β splitting factors, whereas the observed value is 2·8).

(iii) The quantitative predictions of the theory of configurational interaction can be checked in detail for both the attached protons and the C^{13} atoms in the rings.

This same type of analysis has been applied to quite a number of other aromatic negative ions including anthracene, perylene, tetracene, coronene and triphenylene. In each case the spin densities on the different carbon atoms can be evaluated from the simple Hückel theory[20], and then compared with those deduced from the observed hyperfine splittings. A few more examples will serve to illustrate the kind of method adopted.

The anthracene negative ion has been studied in detail by DE BOER[23], and the observed hyperfine pattern is shown in *Figure 40(a)*, together with the predicted spin density distribution from Hückel's treatment, in *Figure 40(b)*. It can be seen from the theoretical figures that there are four different kinds of proton position, and that the position with largest spin density has nearly exactly twice that of the next largest, which has, in turn, twice that

L   145

of the third largest. The fourth position has very little spin density at all, and there are thus three types of interacting protons, with splitting factors in the ratio of 4 : 2 : 1. A build-up of the expected pattern, in a similar way to those already produced, predicts 21 lines, with the two outermost components having only 1 per cent intensity of that of the central peak. The spectrum as shown in *Figure 40(a)* agrees with this predicted pattern very well, although the outermost lines can only just be distinguished above noise.

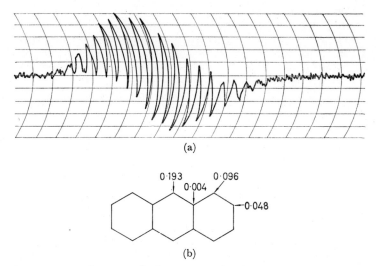

(a)

(b)

Figure 40.   Hyperfine pattern of anthracene negative ion (derivative tracing).   (a) Spectra of negative ion in dimethoxyethane;   (b) Calculated spin densities (After De Boer[23])

The case of the diphenyl negative ion is of some interest as its smaller overall splitting of 21 gauss[23] appears to contradict the figure of 28 gauss predicted by the theory in the last chapter. The observed spectrum and predicted spin densities are shown in *Figure 41(a)* and *(b)* respectively. It will be seen that there is a relatively large spin density on the two central carbon atoms, equal to 24·6 per cent of the total. Since no protons are bound to these positions it follows that there will be no hyperfine contribution from this spin density, and a reduction in the total splitting is therefore to be expected. The measured decrease of 23 per cent is in very good agreement with the predicted figure. This result therefore serves as a further confirmation of the predictions of JARRETT's[24] quantitative calculations on configurational interaction.

The hyperfine patterns of perylene and tetracene negative ions have also been studied[25], and initial observations on the benzene negative ion have been reported[21]. The predicted electron spin densities for the perylene ion[26] are shown in *Figure 42(a)*, and the observed spectrum from the negative ion is shown below in *Figure 42(b)*. The hyperfine pattern is seen to have nine equally spaced groups of lines, and it is also evident from the predicted

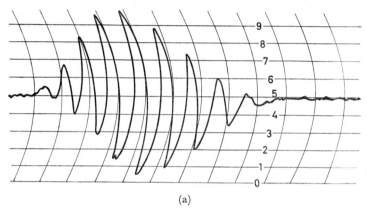

(a)

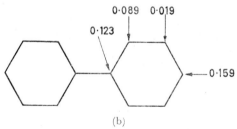

(b)

Figure 41. Hyperfine pattern of diphenyl negative ion. (a) Spectra of negative ion in dimethoxyethane; (b) Calculated spin densities (After De Boer[23])

values that there are eight protons coupled to large and nearly equal spin densities. The unresolved pattern within the groups is due to the other protons attached to carbons with low spin density. The spectrum of *Figure 42(c)* is that of the perylene positive ion and is discussed in more detail below.

5.3.2 *Positive Ions*

Following the results on the aromatic negative ions, it was found by several workers[25,27] that very similar spectra could be obtained from the same aromatic hydrocarbons when dissolved in concentrated sulphuric acid. The similarity between the two types of spectra can be seen in *Figures 42(b)* and *(c)*, the lower spectrum being that from the perylene dissolved in a sulphuric acid. The patterns are seen to have identical shape and components, but differ in their overall splitting.

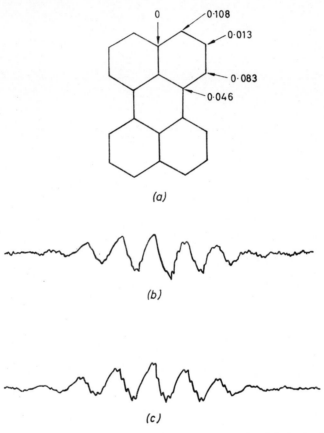

(a)

(b)

(c)

Figure 42. Hyperfine pattern of perylene negative ion. (a) Calculated spin densities; (b) Spectrum of negative ion; (c) Spectrum of postulated positive ion (After Weissman, De Boer and Conradi[25])

These same features are found for all the other aromatic hydrocarbons—i.e. that a spectrum of identical pattern and shape is obtained from the sulphuric acid solution, but with a larger overall splitting. Thus the naphthalene negative ion has an overall splitting of 26 gauss, its sulphuric acid solution one of 28·7 gauss; the anthracene negative ion has an extreme splitting of 26·1 gauss, its sulphuric acid solution one of 31·5 gauss; the tetracene negative ion has one of 24·7 gauss, its sulphuric acid solution one of 29·0 gauss.

As a result of this similarity of spectra WEISSMAN, DE BOER and CONRADI[25] suggested that the hydrocarbons formed positive ions when in the sulphuric acid solution. The molecular orbital theory applicable to them would then be that of a hole in a closed shell, and very similar spin densities[28, 29] and hyperfine coupling constants would thus be expected as for the negative ions. The possibility of such positive ions in hydrocarbons had previously been postulated by WEISS[30], and KAINER and HAUSSER[31] have isolated paramagnetic salts formed in the reaction between aromatic amines and pentachlorides. It was found that the action of antimony pentachloride with the hydrocarbons gave a paramagnetic precipitate of the same colour as the solution of the hydrocarbon in sulphuric acid. Moreover solutions of this precipitate in phosphorus oxychloride had resonance patterns extending over approximately the same range as those in sulphuric acid, although the hyperfine structure was not very well resolved.

Earlier work on other compounds had also shown that free radicals could be produced in concentrated sulphuric acid solution. Thus HIRSHON, GARDNER and FRAENKEL[32] obtained resonance spectra from solutions of bianthrone in sulphuric acid, and a large number of other compounds such as fluorenone, anthraquinone, thiophenol, thiocresol, thio-naphthol and diphenyl sulphide also give spectra[32, 33]. Some of these, such as bianthrone and fluorenone will also give an electron resonance if treated with aluminium chloride in anhydrous ether. It was suggested[32] that positive ions may be formed in these cases, as well, and for the case of bianthrone in sulphuric acid a proton may add to an oxygen atom, and the resultant positive ion could then exist in a diradical state. I.e.

$$^+H + O = (C_{14}H_8) = (C_{14}H_8) = O \longrightarrow HO^+(C_{14}H_8) = (C_{14}H_8) = O$$
$$\longrightarrow H\dot{O}^+ - (C_{14}H_8) - (C_{14}H_8) - \dot{O} \qquad \ldots . (5.2)$$

Later work by MACLEAN and VAN DER WAALS[34] suggests that the observed spectra may not be due to positive ions, however, but rather to oxidation products formed in the sulphuric acid. They carried out a series of measurements on anthracene, pyrene and thianthrene as dissolved in 96 per cent sulphuric acid or in anhydrous hydrofluoric acid. The HF was chosen because it has about the same acid strength as the H_2SO_4 but no oxidizing properties. From the measured basicities of the aromatic compounds it was concluded that they would be present in HF in a protonated form, with no biradical character. If the spectra observed in the sulphuric acid solutions were due to a simple positive ion, the same spectra should be obtained from the hydrofluoric acid solutions. In fact no spectra were observed at all in the latter, although considerably higher concentrations were used than in the sulphuric acid solutions. They therefore concluded that the spectra obtained from sulphuric acid solutions were due to oxidation products and not to univalent positive ions.

It is obvious that more detailed work will have to be carried out on these solutions before the precise radical-forming mechanisms are clear[80,81]. It is certain, however, that whatever the complex formed, it must have a structure very similar to the corresponding negative ion.

5.3.3 *Summary*

In summarizing the results on the aromatic ions, as a whole, it can be seen that they have provided some very well resolved and informative hyperfine patterns, and enabled a detailed comparison to be made between theory and experiment. The development of the theory of configurational interaction has been traced in detail in the last chapter. It was seen there that the constancy of the extreme splitting of the aromatic hyperfine patterns could be explained quantitatively by the predicted spin density interaction parameter of $Q=28$ gauss. The apparent discrepancy obtained with perinaphthene[35] where an extreme splitting of 49 gauss was observed, then lead to the development of the negative spin density concept[36], and with the inclusion of this, the theory now appears to account satisfactorily for all the known data.

A study of the electron resonance of the aromatic negative ions can also be used to obtain other data than the detailed molecular orbitals. Thus it is possible to establish a scale of electronegativity by adding one aromatic hydrocarbon to a solution of the other and determining how the equilibrium is established between them from the intensities of the observed spectra. Anthracene added to naphthalene-ion solution displaces the equilibrium strongly in favour of itself, and the scale of electronegativity in hydrofuran can be written as diphenyl, phenanthrene, naphthalene, anthracene and naphthacene[37]. Information on the rate of electron transfer between the neutral molecule and its negative ion can also be obtained from the widths of the resonance lines. Thus if τ is the lifetime of the radical and it is limited by electron-transfer, a line width of $\Delta\omega \approx \dfrac{1}{\tau}$ will be obtained. Using this fact WARD and WEISSMAN[38] obtained a second-order rate constant of 10^6 l./g.-atom/sec for exchange with the naphthalene negative ion at 30° C.

It should be added that aromatic hydrocarbons are not the only compounds that will give negative ions in solution. Thus electron resonance spectra have also been observed from nitrobenzene, m-dinitrobenzene, and trinitrobenzene when treated with sodium in dimethoxyethane[17]. In these cases the hyperfine interaction with the nitrogen nuclei must also be taken into account[82].

5.4 SEMIQUINONES

Semiquinones are compounds which are formed during the slow reduction of quinones and related compounds. They were first studied magnetically by MICHAELIS et al.[39-41] who showed that the reduction was accompanied by a diminution of the measured diamagnetic susceptibility, followed by a rise to a steady final value. It was therefore postulated that the reduction was a two electron process which took place via a one-electron intermediate. The process can be carried out in reverse, and the hydroquinone, which is the final product of the reduction, can be oxidized back to the quinone via the intermediate semiquinone. The whole reaction system can therefore be represented as

The quinone is usually reduced by the action of glucose in pyridine, while the oxidation of the hydroquinone by atmospheric oxygen in alcoholic solutions can be slowed down if the solution is made slightly alkaline. The intermediate semiquinones are more stable in basic than in acid solution because of the formation of the symmetrical semiquinone ion as shown at the bottom of the reaction illustrated above. Most of the studies on semiquinones have therefore been made on alkaline alcoholic solutions, when a relatively long life-time is obtained for the semiquinone free radical. Alternatively a constant flow process can be employed to maintain a constant radical concentration, such as in the study of the auto-oxidation of hydroquinone[42].

The semiquinone illustrated in the above reaction diagram represents the basic unit taking place in the oxidation-reduction process, but a very large variety of different groups may be substituted on to the edge of the ring to produce different benzo-semiquinone derivatives. A considerable number of these have been studied in detail by electron resonance, and their hyperfine patterns have also proved a fruitful field for testing quantitative theories of hyperconjugation. The spectra of some of these derivatives are discussed briefly below to illustrate these points.

5.4.1 *Benzosemiquinone*

The structure of p-benzosemiquinone is shown in *Figure 43(a)* together with the observed hyperfine pattern[42]. The molecule contains four equivalent protons, with equal spin density on each of their adjacent carbon atoms, and an $(n+1)=5$ pattern is therefore expected with an intensity ratio of $1:4:6:4:1$. This is seen to be exactly the spectrum observed experimentally, and this observation was the first direct and unambiguous detection of the rather unstable p-benzosemiquinone ion in solution.

5.4.2 *Tolu-p-benzosemiquinone*

The structure of the tolu-p-benzosemiquinone (the monomethyl derivative) is shown in *Figure 43(b)* together with its spectrum under low resolution[43]. It is seen that seven equidistant hyperfine components are obtained with an approximately binomial distribution in their intensities. This indicates that there must be equal coupling with six protons, and hence that the hyperfine interaction with the protons of the methyl group must be very similar to that with the protons attached directly to the ring. The explanation of this somewhat surprising fact is the hyperconjugation which mixes the $1s$ orbits of the protons directly with the π orbitals of the ring. This effect has been considered in detail in the last chapter, and the spectrum is striking evidence for the reality of the hyperconjugative type of mixing, and even quantitative agreement with the theory is good[44].

The first three lines of the same spectrum are shown under higher resolution in *Figure 43(c)*, and it is apparent that the lines are not exactly single, and hence that the coupling to all six protons is not exactly equivalent. An analysis of this spectrum in detail[43] shows that whereas the three methyl protons are equivalent, two of the

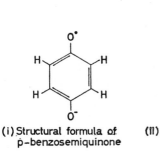

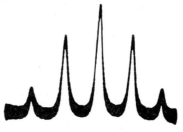

(i) Structural formula of p-benzosemiquinone

(ii) Observed absorption hyperfine pattern

(a)

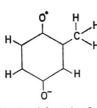

(i) Structural formula of tolu-p-benzosemiquinone

(ii) Observed absorption pattern under low resolution

(b)

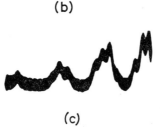

(c)

Figure 43. Spectrum of p-benzosemiquinones. (a) Structural formula and observed absorption pattern of unsubstituted p-benzosemiquinone; (b) Structural formula and observed absorption pattern of tolu-p-benzosemiquinone under low resolution; (c) The first three lines of the tolu-p-benzosemiquinone spectrum under high resolution (After Venkataraman and Fraenkel[42,43])

ring protons have a different coupling to the third. Later higher-resolution studies[45] showed in fact that none of the three ring protons were exactly equivalent. The detail available in such high resolution studies as these shows what precise experimental data are now available for molecular orbital theory.

5.4.3 Other methyl-substituted benzosemiquinones

The spectra of 2, 5-dimethyl-p-benzosemiquinone and 2, 6-dimethyl-p-benzosemiquinone have both been observed[43], and each contains nine major lines split into twenty-one incompletely resolved components. These patterns can be readily interpreted by assuming that the hyperfine interaction is with two equivalent protons on the ring (with interaction constant a), and with six equivalent protons on the two methyl groups (interaction constant b); and that b is somewhat larger than a.

The tetramethyl p-benzosemiquinone gives a well-resolved thirteen-line spectrum[43] as would be expected from the interaction with the twelve equivalent protons of the four CH_3 groups. The interaction constants for these and other derivatives are summarized in *Table 5.1*.

Table 5.1. Interaction Constants for Semiquinones

Derivative of p-benzosemiquinone ion	Interaction Constants (gauss)		
	Ring protons *a*	Methyl protons *b*	Methyl protons calculated[44]
Unsubstituted	2·37	—	—
monomethyl	2·53 and 1·76	2·06	2·00
2.3. dimethyl	2·59	1·71	1·69
2.5. dimethyl	1.83	2.26	2.33
2.6. dimethyl	1·92	2·14	2·00
trimethyl	1·85	1·85 and 2·27	1·70 and 2·34
tetramethyl	—	1·92	2·03
2.5. dichloro-	2·00	—	—
trichloro-	2·11	—	—
2.5. difluoro-	1·3	0·3 (Ring fluorines)	
2.5. dibutyl-	2·13	0·06 (Butyl protons)	

5.4.4 Chloro-substituted benzosemiquinones

Another interesting set of substituted benzosemiquinones are the chloro-derivatives. Each of the protons on the benzosemiquinone can be substituted by a chlorine, and monochloro-, dichloro-, trichloro- and tetrachloro- derivatives thus obtained. The chlorine nucleus has a magnetic moment, but this is $3\frac{1}{2}$ times smaller than that of the proton. It also has a spin of $^3/_2$ instead of $^1/_2$ and hence the splitting due to chlorine atom is only about one tenth of that of a proton with the same unpaired spin density. As a result no observable splitting or structure is produced on the spectra due to the chlorine atoms, and the interaction is therefore only with 3, 2, 1 or 0 protons on the ring.

The derivative tracings of the spectra observed[46] for the four chloro-substituted semiquinones are shown in *Figure 44*, and it is seen

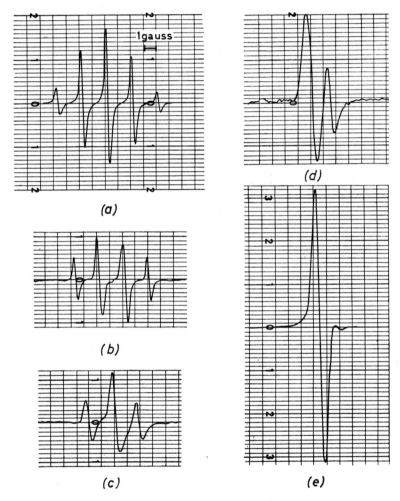

Figure 44. Spectra of chloro-p-benzosemiquinones (derivative tracings). (a) *Unsubstituted benzosemiquinone;* (b) *Monochlorosemiquinone;* (c) *2.3-di-chlorosemiquinone;* (d) *Trichlorosemiquinone;* (e) *Tetrachlorosemiquinone (After Wertz and Vivo[46])*

that the prediction of $(n+1)$ hyperfine components is confirmed throughout. The widths of the component lines is less than 0·4 gauss, and no trace of structure is seen, showing that the interaction with the chlorines must be very small.

Various fluorine- substituted p-benzosemiquinones have also been examined[47]. The fluorine nucleus has a very much larger magnetic moment than chlorine and a spin of $1/2$, and hence hyperfine splitting due to this should be of the same order as that from the proton. The spectra obtained from the fluoro- derivatives are in general considerably more complex than those observed for the unsubstituted compounds, but the monofluoro- and 2, 5-difluoro-p-benzosemiquinone spectra have been interpreted in terms of a proton splitting of 1·1 gauss and a fluorine splitting of 0·3 gauss.

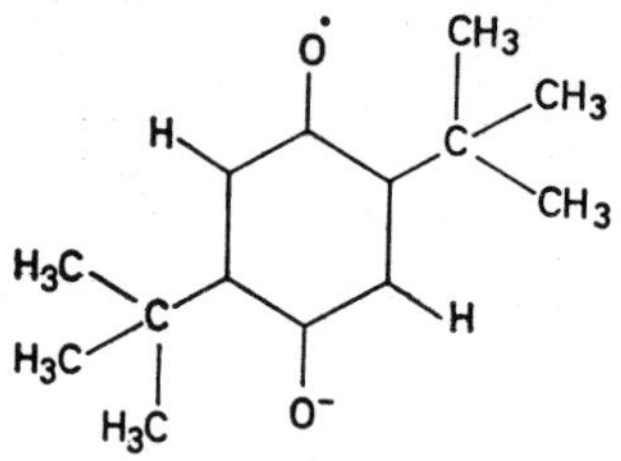

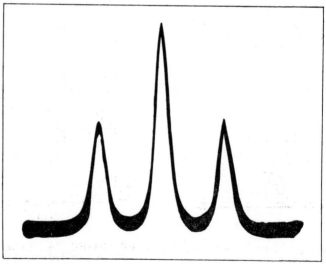

(a)

(b)

156

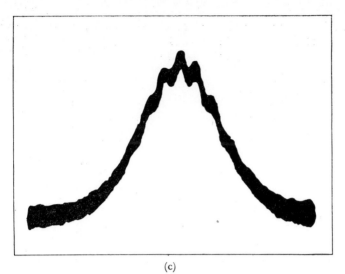

(c)

Figure 45. Spectra of 2, 5-di-butyl-benzosemiquinone. (a) Structural formula; (b) Absorption spectrum from concentrated solutions; (c) Central line under high resolution in dilute solution (After Fraenkel[45])

It is interesting that the proton interaction has been reduced to under half the value of 2·4 gauss found for the unsubstituted case, and is appreciably less than that for the dimethyl derivatives.

5.4.5 Other semiquinones

A large number of other semiquinones have now been studied by various workers, including o-benzosemiquinones[48], and some naphthosemiquinones[49]. In most cases the hyperfine pattern can be analysed in the manner outlined in the cases above, and detailed information obtained on the spin density distribution. Two cases are of particular interest, phenyl-p-benzosemiquinone and dibutyl-p-benzosemiquinone. In the phenyl derivative a whole benzene ring has been substituted in place of one edge proton, but no extra splitting due to the protons of the phenyl group was detected in the hyperfine pattern[33]. This shows that the unpaired electron spin density does not couple appreciably with the additional benzene ring.

In the 2, 5-di-t-butyl-p-benzosemiquinone two $C(CH_3)_3$ groups are substituted on to the ring as shown in Figure 45(a). The spectrum obtained from relatively concentrated solutions[45] is shown in Figure 45(b) and the triplet is due to equal hyperfine interaction with the two ring protons. The appearance of the central line when the

157

concentration of the solution is reduced is shown in *Figure 45(c)*, and it can be seen that there is a small hyperfine interaction with all the protons of the two butyl groups. Equal coupling with the eighteen protons should produce nineteen lines with the central line about $50,000$, $\left(\dfrac{18!}{(9!)^2}\right)$, times as intense as the extreme components. It is not surprising therefore that only eleven of the nineteen have been observed[45]. The separation between these butyl components is about 60 milligauss, and they are a good illustration of the high resolution required for some electron resonance studies.

In summarizing the work on the semiquinones it can be seen that they have formed an ideal system for studying the effect of different substitutions, and, in this sense, are more flexible than the hydro-carbon negative ions. In the same way that the hydrocarbon ions were used to confirm the theories of configurational interaction, so the results on the substituted semiquinones have confirmed the predictions of direct spatial overlap, or hyperconjugation. The quantitative agreement between the measured values of hyperfine splitting and those predicted by molecular orbital theory is also good. The theory as applied to the semiquinones has been given by BERSOHN[50] and McCONNELL[51], and, as an example, predicts a spin density of 0·078 for the four carbon atoms of *p*-benzosemiquinone, whereas the measured value is 0·07.

It should be mentioned that most of the semiquinone spectra decay gradually, and somewhat change their form in the process. It is possible to follow these changes[52], even if quite rapid, in the electron resonance spectrometer and analyse the chemical reactions. Experiments by BIJL, KAINER and ROSE-INNES[53] have also shown that they can be stabilized by absorption on basic surfaces such as alumina and silica-gel when their resonance spectrum is obtained without diminution. It might also be added that other types of radical can also be formed by oxidation in alkaline alcoholic solutions, such as those derived from diarylamines where a large triplet splitting of the hyperfine pattern by interaction with the nitrogen nucleus is obtained[54].

Another series of radical ions which are produced by partial oxidation are the Würster's blue derivatives, which are stabilized in acid solution. These have the general formula of and quite a large number of different derivatives have been studied[55,56]. All the spectra have very complex hyperfine patterns, often with marked triplet groupings from an interaction with the nitrogen atoms.

5.5 TRIPHENYLMETHYL, DIMESITYLMETHYL AND PERINAPHTHENE

These three radicals have been grouped together for consideration since their spectra have larger extreme splittings than normal and appear to contradict the simple configurational interaction theory at first sight.

5.5.1 *Triphenylmethyl*

The structural formula for triphenylmethyl and its spectrum at a dilution of 10^{-3} molar in benzene, and in high homogeneity magnetic field[57], is shown in *Figure 46(a)*. On a first analysis this spectrum consists of 21 major peaks with groups of about four closely-spaced sub-components. It may be noted that a total of 196 hyperfine lines would be expected if the proton positions at all the ortho-, para- and meta- positions of the three phenyl rings were equivalent. A complete detailed interpretation of this spectrum has yet to be given, and the complexity involved in such an analysis is only too obvious. Detailed information on the spin density at the central methyl carbon atom has been obtained by C^{13} substitution, however. WEISSMAN and SOWDEN[58] prepared a solution with 53 per cent C^{13} in the methyl position, and obtained a doublet splitting superimposed on the singlet from the normal C^{12}. (If the concentration of the solution is greater than 1·0 molar, or if the triphenylmethyl is dissolved in hexaphenylethane, the hyperfine pattern disappears and a single line of 5 gauss width is obtained.) The splitting between the doublet lines was 22 gauss and this indicates appreciable spin density at the central carbon atom.

The fact that an appreciable spin density was present at the methyl carbon, which produces no hyperfine interaction, and that the observed pattern still had an extreme splitting of 25 gauss, at first seemed inconsistent[35] with the total splitting parameter of 28 gauss. This splitting parameter is that derived by the general theory of configurational interaction in aromatics. This inconsistency was removed with the introduction of the effect of negative spin densities[59], however, as discussed in the last chapter. The valence bond calculations of BROVETTO and FERRONI[60] show that the meta- spin densities are negative and have a magnitude 35 per cent as great as the sum of the positive spin densities on the ortho- and para- carbon atoms. It would therefore appear that once the spin density sign has been taken into account, the generalized splitting parameter still holds.

5.5.2 Dimesitylmethyl

The structural formula and the observed hyperfine pattern of dimesitylmethyl, as obtained by JARRETT and SLOAN[57], are shown in *Figure 46(b)*. It is seen that this radical now has a proton attached to the central methyl carbon, and any large spin density on the carbon atom will therefore produce a large doublet hyperfine splitting. Such a splitting is seen to form the main feature of the

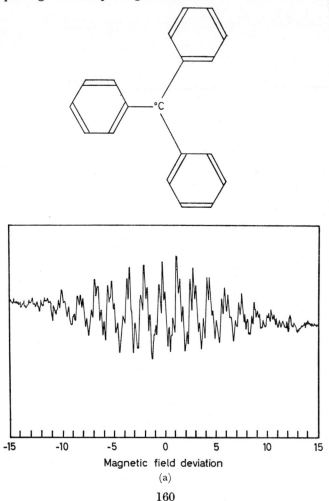

Magnetic field deviation

(a)

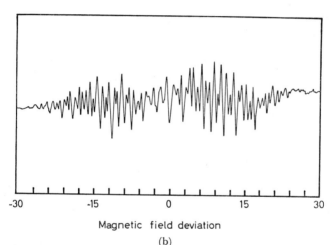

Magnetic field deviation

(b)

Figure 46. Spectra of triphenylmethyl and dimesitylmethyl. (a) Structural formula, and observed spectrum of triphenylmethyl in dilute benzene solution (derivative tracing); (b) Structural formula, and observed spectrum of dimesitylmethyl (derivative tracing) (After Jarrett and Sloan[57])

observed spectrum, as the lines fall into two groups of 35 components, with a mean separation of 24 gauss. The thirty-five components are explained by a strong interaction with the four protons of the metapositions, thus producing a five-line spectrum on each doublet, followed by a weaker interaction with the six protons of the para-methyl groups, which splits each of the five lines into seven. It may be shown by molecular orbital treatment[57] that if the radical is nearly planar, hyperconjugation interaction will be 400 times greater for the para-methyl groups than for the ortho-methyl groups.

This spectrum affords a very good example of how a complex pattern can be analysed quite quickly into its main components, and these then related to the spin densities round the radical. It is also seen that the large splitting due to the spin density on the central carbon is in good agreement with the C^{13} measurements on triphenylmethyl[58], and that the concept of negative spin densities[59] must also be included in the theoretical explanation of the large overall splitting.

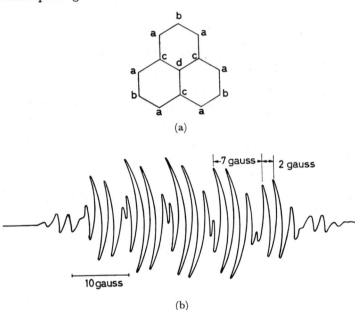

(a)

(b)

Figure 47. Spectrum of perinaphthene. (a) *Structure with theoretical values*[79] *of spin densities given by* $\rho_a = +0.32$; $\rho_b = -0.18$; $\rho_c = -0.22$; $\rho_d = +0.26$; (b) *Derivative tracing of hyperfine splitting*

5.5.3 Perinaphthene

SOGO, NAKAZAKI and CALVIN[35] found that, after standing for several hours, a solution of perinaphthene in carbon tetrachloride became yellow in colour and would give an absorption in an electron resonance spectrometer. The spectrum that they observed is shown in *Figure 47* together with the structure of the perinaphthene radical. The spectrum is seen to consist of seven lines with an approximate intensity ratio of $1:6:15:20:15:6:1$ with each line further split into a quadruplet. This pattern can be readily explained if it is assumed that a symmetrical free radical is produced from the perinaphthene by the removal of a hydrogen atom. There are then six symmetrically located proton positions (*a* in the figure), with high spin density, which produce the seven main components, and the other three symmetrically related protons, *b*, then split each of these into four. The two splittings are 7·3 and 2·2 gauss respectively, and this ratio can be predicted by second order valence bond theory.

The overall splitting of the spectrum is 49 gauss, and it was this large splitting that first led to the suggestion[35] that the generalized treatment of configurational interaction was not always applicable. Recent valence bond calculations[79] on this radical have shown, however, that the spin densities are approximately $\rho_a = +0\cdot32$, $\rho_b = -0\cdot18$ and that the larger splitting can therefore again be explained by the occurrence of the negative spin densities. There is, in fact, reasonable agreement between these calculated spin densities and those obtained experimentally using a Q value of 28 gauss, when $|\rho_a|$ is obtained[79] as 0·325, and $|\rho_b|$ as 0·098.

5.6 Radicals Containing Sulphur

Free radicals containing sulphur are an interesting group because *g*-values quite appreciably different from 2·0 are often obtained, in contrast to nearly all the other free radicals investigated. Electron resonance absorption from such compounds as thiophenol, thiocresol, thionaphthol, and diphenyl disulphide when dissolved in concentrated sulphuric acid were observed by Hirshon, Gardner and Fraenkel[32]; and Wertz and Vivo[33] obtained two distinct hyperfine patterns from dilute solutions of these in concentrated sulphuric acid. The observed spectrum of *p*-thiocresol is shown in *Figure 48*, and the separation between the two groups of lines is about 10 gauss. The *g*-values for the centres are 2·0151 and 2·0081, with respect to a 1,1-diphenyl-2-picryl hydrazyl *g*-value of 2·0036. The fact that these two hyperfine groups arise from two separate radicals is shown by the gradual disappearance of the low-field set on standing, while the high-field set remain observable for over a month. It was suggested that the more stable high-field set, with a *g*-value close to the free spin value, were due to thianthrene, and that the five hyperfine components originated from the four-proton system of one of the thianthrene rings, as in benzosemiquinone. This was confirmed by the fact that a similar set of two groups was obtained from solutions of thiophenol and thianthrene itself. It would therefore appear that the hyperfine group with a *g*-value of 2·015 is due to a radical produced by oxidation of a complex containing sulphur. A possible formula for this has been suggested[33] as $(C_6H_5S\cdot)\,H^+$, with the unpaired electron strongly localized on the sulphur atom. It may be noted that strong localization will generally produce larger spin-orbit coupling, and hence an appreciable shift of the g-value from 2·0023.

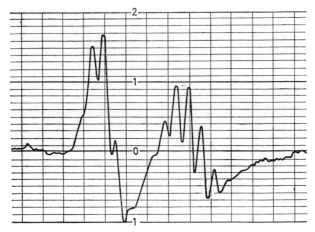

Figure 48.   Spectrum obtained from p-thiocresol *in concentrated sulphuric acid (After Wertz and Vivo*[33]*)*

GARDNER and FRAENKEL[61] have also studied the electron resonance obtained from liquid sulphur. This has a g-value of 2·024 and is postulated as due to the breakage of S-S bonds in polymeric chains. They also reported the occurrence of two lines when sulphur was dissolved in oleum. These two lines have been studied in more detail by INGRAM and SYMONS[62], and if the solutions are frozen the lines take on a marked asymmetrical shape, indicative of a g-value anisotropy. When the sulphur was dissolved in concentrated oleum (65 per cent SO_3) only one line was obtained, with a g-value varying from 2·003 to 2·018, whereas when dissolved in dilute oleum (20 per cent. SO_3) another asymmetrical line appeared, with a g-value varying from 2·025 to 2·032. The fact that these were genuine g-value variations, and not unresolved hyperfine or electronic splittings, was confirmed by repeating the measurements at 1·25 cm wavelength. The change in the relative intensity of the two lines with concentration of oleum showed that two different radical species were again being observed. The mean g-value of the line appearing in the diluted oleum, just before solid sulphur separates out, is the same as that observed in the liquid sulphur[61], and it is therefore suggested[62] that this also is due to sulphur in a polymeric form.

These results on the sulphur-containing radicals are of interest in showing that considerable localization of the unpaired electron, with appreciable spin-orbit coupling and relatively large g-value shifts, can occur in some cases. Only in radicals containing sulphur or oxygen has this been observed to any noticeable extent to date, and this is almost certainly to be associated with the high electron affinity of these two atoms. It should be noted, from the experimental point of view, that it is advisable to have two spectrometers available at somewhat different wavelengths if sulphur or oxygen-containing radicals are to be studied. It is then possible to establish unambiguously whether an asymmetrical line is due to a g-value spread or to unresolved structure from more than one radical. In the former case the measured width of the line will increase linearly with the frequency of measurement, whereas in the latter case it will remain constant.

5.7 OTHER STABLE RADICALS

A large number of other stable free radicals have been observed by electron resonance, and a list of some of these is given in *Table* 5.2. This is not a comprehensive nor a complete list since new results are constantly being reported, but it serves to illustrate the variety of compounds that can be studied by this means. The analysis of the observed hyperfine patterns proceeds in a similar way to the cases outlined in the previous sections and the spectra show that, in most cases, the unpaired electrons are moving in highly delocalized orbitals.

One radical of particular interest is the peroxylamine disulphonate ion, $(SO_3)_2 NO^{2-}$, since its hyperfine splitting has been traced back in detail to zero magnetic field. The spectrum consists of three single lines with a splitting of 13 gauss when observed in an X-band spectrometer[68], and this splitting is due to an interaction with the single nitrogen atom of spin $I=1$. This forms a particularly simple system to study and TOWNSEND, WEISSMAN and PAKE[69] have traced the splitting back into zero field to see if the BREIT-RABI[70] formulae for the Zeeman effect would apply. These formulae are those derived for the hyperfine interaction between the nucleus and a single electron of a free atom. The splitting of 13 gauss, as measured at high fields corresponds to 36 Mc/s, and if the resonance frequency is reduced towards this value, the simple picture of hyperfine splitting of the two electronic levels breaks down. The levels are then defined by a quantum number F which couples I and S, and as zero-field is approached the energy levels diverge widely from a linear plot. TOWNSEND, WEISSMAN and

P<small>AKE</small>[69] took a series of measurements at frequencies from 9·2 Mc/s up to 120 Mc/s with d.c. magnetic fields up to 50 gauss, and were able to show that the energy level system accurately obeyed the B<small>REIT</small>-R<small>ABI</small> formulas[70] with a splitting parameter of $\Delta v = 54\cdot7$ Mc/s. It is interesting to see that the more complex case of an electron in the molecular orbital of a free radical can still be explained by this treatment. L<small>LOYD</small> and P<small>AKE</small>[71] have also studied the saturation of the peroxylamine disulphonate spectra at low fields in order to establish the relaxation mechanism in such absorptions as these.

References are included in *Table 5.2* so that the more detailed work on any of the radicals can be consulted. It should be noted that the stable radicals formed during polymerization processes, or during the charring or coalification of organic matter, are considered in detail in Chapter 7.

Table 5.2. Stable Organic Radicals Studied by Electron Resonance

Free Radical	Line Width* or extreme h.f.s. splitting (gauss)	Reference
I *IN THE SOLID STATE*		
Bianthrone (heated)	10	73
Dianisyl nitric oxide	30	72
1,1-Diphenyl-2-picryl hydrazyl	1·9 (single crystal)	1, 12
Diphenylenetriphenylethyl	5	17
Kenyon-Banfield radical	8·9	72, 17
Pentaphenylcyclopentadienyl		74
Picryl-aminocarbazyl	0·5 (single crystal)	14
Porphyrexide	17·0	72
Porphyrindene	10·7	17
Tetramethyl benzidine formate	3·4	17
Tetraphenylstibonium peroxylamine disulphonate	100	17
Tri-*p*-xenylmethyl	5·7	17
Tri-*t*-butyl phenoxyl	7·7	75
Tri-phenylamine perchlorate	2	17
Tri-*p*-anisylaminium perchlorate	0·68	77
Tri-*p*-aminophenylaminium perchlorate	0·33	77
Tri-*p*-nitrophenylmethyl	0·7	17
Violanthrene	26	63
Violanthrone	30	63
Würster's blue perchlorate	2·7	17
Products with potassium in dioxane		
Anthracene	3·7	65
Benzophenone	3·8	65
Chrysene	2·2	
Naphthalene	5	65
Phenanthrene	3·2	65
Pyrene	2·5	
Triphenyl benzene	13	65

* The full width between half-power points

II. *IN SOLUTION*	Line Width* or Extreme h.f.s. splitting (gauss)	Number of h.f.s. components	Reference
(a) *Aromatic Ions*			
Anthracene	26	21	23
Benzene	22·5	7	21
m-Dinitrobenzene	25	8	56, 17
Diphenyl	21	9	23
Naphthalene	27	17	19
Naphthalene-C^{13}	34·3	34	22
Naphthalene-deuterated	29	15 or 16	19, 21
Nitrobenzene	25	10	17
Perinaphthene	49	7×4	35
Perylene	24	9	25
Tetracene	25	31	25
Trinitrobenzene	25	8	17
(b) *Semiquinones*			
(i) p-Benzosemiquinone	9·48	5	42
mono-methyl- ,,	14	7+(further structure)	43
tetra-methyl ,,	23	13	43
mono-chloro- ,,	6·0	4	46
tri-chloro- ,,	2·11	2	45
tetra-chloro- ,,	0·4	1	46
2, 5-di-t-butyl- ,,	4·3	3×19	45
(ii) 1, 4.-Naphthosemiquinone	8	3×5	49
2, 3.-dimethyl- ,,	12	7×5	49
(iii) o-Benzosemiquinone	10	3×3	66
4-tert-butyl- ,,	6	2×11	66
3-phenyl- ,,	8	7	66
(iv) oxidized 1, 2, 3.-Benzenetriol	7	2×3	67
(c) *Triarylaminium perchlorates*			
tri-anisyl	20	3	64
tri-diphenyl	20	3	64
(d) *Würster's Salts* (positive ions)			
Unsubstituted	28	15	55
N methyl-	46	24	55
NN dimethyl-	73	27	55
NN' dimethyl	76	9×3	55
NN dimethyl- - N'N' dideutero-	51	7×3	55
tetramethyl-	88	13×3	55
(e) *Miscellaneous*			
Carbazyl (Picryl-amino carbazyl)	60	7	13, 14
Difluorenyl nitrogen	9	10	16
Dimesitylmethyl	48	2×35	57
Di-p-anisyl nitricoxide	14	—	76
Diphenyl-dinitro-sulphonyl-hydrazyl	60	9	13
1, 1.-Diphenyl-2-picryl hydrazyl	58	5	12
Peroxylamine disulphonate	26	3	68
Phenazine	60	5	78
Sodium trimesityl boron	42	4	16
Triphenylmethyl	25	21×4	57

* The full width between half-power points

REFERENCES

[1] HOLDEN, A. N., KITTEL, C., MERRITT, F. R. and YAGER, W. A. *Phys. Rev.* 77 (1950) 147
TOWNES, C. H. and TURKEVITCH, J. *Ibid.* 77 (1950) 148

[2] GOLDSMICHT, S. and RENN, K. *Chem. Ber.* 55 (1922) 636

[3] PAKE, G. E., WEISSMAN, S. I. and TOWNSEND, J. *Disc. Faraday Soc.* 19 (1955) 147

[4] COHEN, V. W., KIKUCHI, C. and TURKEVICH, J. *Phys. Rev.* 85 (1952) 379

[5] KIKUCHI, C. and COHEN, V. W. *Phys. Rev.* 93 (1954) 394

[6] SINGER, L. S. and KIKUCHI, C. *J. chem. Phys.* 23 (1955) 1738

[7] BLOEMBERGEN, N. and WANG. *Phys. Rev.* 93 (1954) 72

[8] LLOYD, J. P. and PAKE, G. E. *Ibid.* 92 (1953) 1576

[9] LIVINGSTON, R. and ZELDES, H. *J. chem. Phys.* 24 (1956) 170

[10] GARSTENS, M. A. *Phys. Rev.* 93 (1954) 1228
— SINGER, L. S. and RYAN, A. H. *Phys. Rev.* 96 (1954) 53

[11] BECKER, S. *Ibid.* 99 (1955) 1681

[12] HUTCHISON, C. A., PASTOR, R. C. and KOWALSKY, A. G. *J. chem. Phys.* 20 (1952) 534

[13] JARRETT, H. S. *Ibid.* 21 (1953) 761

[14] KIKUCHI, C. and COHEN, V. W. *Phys. Rev.* 93 (1954) 394

[15] BELJERS, H. G., VAN DER KINT, L. and VAN WIERINGEN, J. S. *Ibid.* 95 (1954) 1683

[16] WEISSMAN, S. I., TOWNSEND, J., PAUL D. E. and PAKE, G. E. *J. chem. Phys.* 21 (1953) 2227

[17] CHU, T. L., PAKE, G. E., PAUL, D. E., TOWNSEND, J. and WEISSMAN, S. I. *J. Phys. Chem.* 57 (1953) 504

[18] LIPKIN, D., PAUL, D. E., TOWNSEND, J. and WEISSMAN, S. I. *Science*, 117 (1953) 534

[19] TUTTLE, T. R., WARD, R. L. and WEISSMAN, S. I. *J. chem. Phys.* 25 (1956) 189

[20] HÜCKEL, E. *Z. Phys.* 70 (1931) 204

[21] WEISSMAN, S. I., TUTTLE, T. R. and DE BOER, E. *J. Phys. Chem.* 61 (1957) 28

[22] TUTTLE, T. R. and WEISSMAN, S. I. *J. Chem. Phys.* 25 (1957) 190

[23] DE BOER, E. *Ibid.* 25 (1956) 190

[24] JARRETT, H. S. *Ibid.* 25 (1956) 1289

[25] WEISSMAN, S. I., DE BOER, E. and CONRADI, J. J. *Ibid.* 26 (1957) 963

[26] DE BOER, E. Thesis, University of Amsterdam

[27] YOKOZAWA, Y. and MIYASHITA, I. *J. chem. Phys.* 25 (1956) 796

[28] COULSON, C. A. and RUSHBROOKE, G. S. *Proc. Camb. phil. Soc.* 36 (1940) 1930

[29] POPLE, J. A. *Trans. Faraday Soc.* 49 (1953) 1375

[30] WEISS, J. *Nature, Lond.* 147 (1941) 512

[31] KAINER, H. and HAUSSER, K. *Chem. Ber.* 86 (1933) 1563

[32] HIRSHON, J. M., GARDNER, D. M. and FRAENKEL, G. K. *J. Amer. chem. Soc.* 75 (1953) 4115

[33] WERTZ, J. E. and VIVO, J. L. *J. chem. Phys.* 23 (1955) 2193

[34] MACLEAN, C. and VAN DER WAALS, J. H. *Ibid.* 27 (1957) 827

[35] SOGO, P. B., NAKAZAKI, M. and CALVIN, M. *Ibid.* 26 (1957) 1343

[36] McCONNELL, H. M. and CHESNUT, D. B. *Ibid.* 27 (1957) 984

[37] WERTZ, J. *Chem. Rev.* 55 (1955) 922

[38] WARD, R. L. and WEISSMAN, S. I. *J. Amer. chem. Soc.* 76 (1954) 3612

[39] MICHAELIS, L., SCHUBERT, M. P., REBER, R. K., KUCK, J. A. and GRANICK, S. *Ibid.* 60 (1938) 1678

[40] — *Ann. N.Y. Acad. Sci.* 40 (1940) 39

[41] — GRANICK, S. *J. Amer. Chem. Soc.* 70 (1948) 624, 4275

[42] VENKATARAMAN, B. and FRAENKEL, G. K. *Ibid.* 77 (1955) 2707
[43] — — *J. chem. Phys.* 23 (1955) 588
[44] BERSOHN, R. *Ibid.* 24 (1956) 1066
[45] FRAENKEL, G. K. *Ann. N.Y. Acad. Sci.* 67 (1957) 553
[46] WERTZ, J. E. and VIVO, J. L. *J. chem. Phys.* 23 (1955) 2441
[47] FRANK, P. J. and GUTOWSKY, H. S. In press
[48] HOSKINS, R. H. *J. chem. Phys.* 23 (1955) 1975, 2461
[49] WERTZ, J. E. and VIVO, J. L. *J. chem. Phys.* 24 (1956) 479
[50] BERSOHN, R. *Ibid.* 24 (1956) 1066
[51] MCCONNELL, H. M. *Ibid.* 24 (1956) 632
[52] SCOTT BLOIS *Ibid.* 23 (1955) 1351
[53] BIJL, D., KAINER, H. and ROSE-INNES, A. C. *Nature, Lond.* 174 (1954) 830
[54] HOSKINS, R. *J. chem. Phys.* 25 (1956) 788
[55] WEISSMAN, S. I. *Ibid.* 22 (1954) 1135
[56] PAKE, G. E., WEISSMAN, S. I. and TOWNSEND, J. *Disc. Faraday Soc.* 19 (1955) 147
[57] JARRETT, H. S. and SLOAN, G. J. *J. chem. Phys.* 22 (1954) 1783
[58] WEISSMAN, S. I. and SOWDEN, J. C. *J. Amer. chem. Soc.* 75 (1953) 503
[59] MCCONNELL, H. M. and CHESNUT, D. B. *J. chem. Phys.* 27 (1957) 984; 28 (1958) 107
[60] BROVETTO, P. and FERRONI, S. *Nuovo Cim.* 5 (1957) 142
[61] GARDNER, D. M. and FRAENKEL, G. K. *J. Amer. chem. Soc.* 78 (1956) 3279
[62] INGRAM, D. J. E. and SYMONS, M. C. R. *J. chem. Soc.* (1957) 2437
[63] YOKOZAWA, Y. and TATZSUZAKI, I. *J. chem. Phys.* 22 (1954) 2087
[64] GILLIAM, O. R., WALTER, R. I. and COHEN, V. W. *Ibid.* 23 (1955) 1540
[65] PASTOR, R. C. and TURKEVICH, J. *Ibid.* 23 (1955) 1731
[66] HOSKINS, R. H. *Ibid.* 23 (1955) 1975
[67] — and LOY, B. R. *Ibid.* 23 (1955) 2461
[68] PAKE, G. E., TOWNSEND, J. and WEISSMAN, S. I. *Phys. Rev.* 85 (1952) 682
[69] TOWNSEND, J., WEISSMAN, S. I. and PAKE, G. E. *Ibid.* 89 (1953) 606
[70] BREIT, G. and RABI, I. I. *Ibid.* 38 (1931) 2082
[71] LLOYD, J. P. and PAKE, G. E. *Ibid.* 94 (1954) 579
[72] HOLDEN, A. N., YAGER, W. A. and MERRITT, F. R. *J. chem. Phys.* 19 (1951) 1319
[73] NILSEN, W. G. and FRAENKEL, G. K. *Ibid.* 21 (1953) 1619
[74] KOZYREV. B. M. and SALIKHOV, S. G. *C.R. Acad. Sci. U.R.S.S.* 58 (1947) 1023
[75] WERTZ, J. E. *Chem. Rev.* 55 (1955) 921
[76] HOLDEN, A. N., KITTEL, C., MERRITT, F. R. and YAGER, W. A. *Phys. Rev.* 77 (1950) 147
[77] CODRINGTON, R. S., OLDS, J. D. and TORREY, H. C. *Ibid.* 95 (1954) 607
[78] FELLION, Y. and UEBERSFELD, J. *Arch. Sci. Genève,* 10 (1957) 95
[79] MCCONNELL, H. M. and DEARMAN, H. H. *J. chem. Phys.* 28 (1958) 51
[80] HIRSHBERG, Y. and WEISSMAN, S. I. *Ibid.* 28 (1958) 739
[81] KON, H. and BLOIS, M. S. *Ibid.* 28 (1958) 743
[82] WARD, R. L. and KLEIN, M. P. *Ibid.* 28 (1958) 518

RADICALS PRODUCED BY IRRADIATION

6.1 INTRODUCTION

THE fission of chemical bonds by radiation is one of the most general and direct methods available for producing molecular fragments and free radicals. The bonds between atoms can often be broken by ultra-violet, X- or γ-ray irradiation, and the minimum energy required is determined by the strength of the bond itself. Most bond strengths are of the order of ten electron volts, and thus correspond to wavelengths in the ultra-violet region. Hence the most sensitive measurements of the energy required to form free radicals will be those employing u.v. irradiation, and in this region the energy necessary for the onset of radical production can often be determined directly by simply varying the wavelength of the incident illumination. As an example, it has been found[1] that free benzyl radicals are produced from benzyl chloride on irradiation with 2540 Å but not on irradiation with 3650 Å. It therefore follows that the energy required to form these radicals must be greater than 4 electron volts (92 kg-cals. per mole).

The energies available in X- or γ-ray irradiation are very much higher, from a thousand to a thousand million electron volts. A large number of possible bonds can therefore be broken and most substances in the solid phase, whether amorphous or crystalline, will show abundant free-radical spectra when X- or γ-irradiated. This high energy irradiation therefore has the advantage that high concentrations of free radicals are usually produced, and the electron resonance spectra have high signal-to-noise ratios. It has the disadvantage, however, that it is often difficult to assign a definite mechanism to the method of free radical production, because so many different types of damage are possible. In many cases, overlapping spectra from several different radical species may also result.

In all irradiation experiments performed to date it has been necessary to employ some kind of trapping mechanism in order to accumulate a large enough radical concentration. In principle it should be possible to study the dynamic concentration of radicals in solution during u.v. or X-ray irradiation. The concentration present at any given time would then be determined by a balance between their rate of formation and rate of recombination. The latter rate is usually very high, unless viscous media are employed, and as a result the dynamic concentration of the radicals is often less than 10^{-10} molar and hence too small for detection. Several laboratories are working on this problem at the moment, however, and it should not be long before some successful measurements on dynamic radical concentrations are reported. It should be noted in this connection that the finite lifetime of the radicals may also be a limiting factor in their detection. If their mean lifetime is τ seconds, their absorption line will have a width of not less than $\frac{1}{\tau}$c/s, $\left(\frac{3.10^{-7}}{\tau}\text{gauss}\right)$. For high sensitivity work this extra broadening should not be greater than 1 gauss, and hence the lifetime of the radicals should not be less than a third of a microsecond.

So far, however, all the published work on free radicals produced by irradiation has been concerned with those formed and trapped in the solid or viscous phase. In X- or γ-ray irradiation there is always sufficient kinetic energy imparted to the molecular fragments or broken bonds for them to move away with some speed. They therefore become trapped in the solid at a sufficient distance apart to prevent recombination. As a result radicals can usually be formed and trapped however viscous or rigid the solid phase may be. The conditions are more critical with ultra-violet irradiation, since the resulting fragments or radicals have not so much excess kinetic energy after the bond fission. If the solid is too rigid the radicals will not be able to move apart, and hence will recombine immediately. On the other hand, the solid must be sufficiently rigid to hold the radicals apart after their initial formation. The practical compromise between these two requirements is discussed in detail in the next section.

Since the use of ultra-violet irradiation is the more sensitive tool in this work, the results obtained by this method are discussed first. The measurements employing X- or γ-ray irradiation are then summarized under the two headings of ' organic ' and ' inorganic ' compounds. It should be noted that throughout this chapter the incident radiation is assumed to break the chemical bond completely thus leaving two molecular fragments, each with an unpaired electron. The case of bond excitation, instead of breakage, in which one of the bonding electrons is excited into a higher energy level with reversed spin, is dealt with later in Chapter 8 under the heading of ' the triplet state '.

6.2   ULTRA-VIOLET IRRADIATION

6.2.1   *Trapping Techniques*

The first electron resonance spectra from active radicals formed by ultra-violet irradiation and trapped at low temperatures were obtained by INGRAM, HODGSON, PARKER and REES[1]. In these experiments the compounds to be photolysed were dissolved in various hydrocarbons and the solutions were then frozen at liquid nitrogen temperatures to form rigid glasses. As noted above, the viscosity of the resultant glass is an important parameter of the experiment since the structure must not be too rigid or the radicals will never separate after formation. On the other hand, it must be rigid enough to reform behind the radicals once their recoil has taken place, and so trap them permanently in the solid structure. The use of hydrocarbon glasses at liquid oxygen or nitrogen temperatures is ideal for this purpose since the viscosity of the medium can be altered by either varying the constitution of the hydrocarbon mixture, or by altering the temperature of the specimen. The radicals will recombine, of course, immediately the glass is warmed to its melting point. If the temperature is raised gradually, however, it is possible to observe various stages of radical reaction and decay as the viscosity of the glass slowly decreases. It is possible, in principle, to use room temperature glasses for such a trapping process, and BIJL and ROSE-INNES[2] were able to form stable radicals in plastic films by this means. The viscosity of such ' high temperature ' glasses cannot be varied so precisely, however, and they are not very suited for the study of radical reaction and decay under different conditions.

This low temperature glass trapping technique was used for studying reactive molecules in the early work of LEWIS and LIPKIN[3], and more recently by NORMAN and PORTER[4] who detected and identified the trapped free-radicals by u.v. absorption spectroscopy. Two particular mixtures of hydrocarbon solvents were found to produce good glasses on deep-freezing, the one labelled P.He.H consists of 3 parts of isopentane and 2 parts of cyclohexane, the other labelled E.P.A. consists of 5 parts of ether, 5 parts of isopentane and 2 parts of ethanol. It may be noted, however, that the glasses required for electron resonance work are not quite so critical as those required for u.v. absorption spectroscopy since small imperfections which would badly scatter u.v. wavelengths have no effect on the microwave radiation.

6.2.2 Initial Experiments

In the initial electron resonance experiments[1] a series of different compounds were dissolved in these hydrocarbon mixtures and then irradiated after cooling to liquid nitrogen temperatures. Identical experiments on the pure solvents alone showed that no radicals were produced by u.v. photolysis. Hence any electron resonance spectra that were obtained could be attributed to radicals formed

172

by photolysis of the particular compound dissolved in the hydro-carbon mixture. It was shown in this way[1] that active radicals were formed from ethyl iodide, benzylamine and benzyl chloride when irradiated with 2540 Å wavelength, but not when irradiated with 3650 Å wavelength. On the other hand, active radicals could be produced from hydrogen peroxide in water and various oxalate complexes in ether-alcohol-water mixtures when irradiated by either wavelength. It was also found that the mobilities of the radicals could be studied by this means. Thus the ethyl and benzyl radicals in the hydrocarbon glasses would diffuse and recombine if the temperature were raised from that of liquid nitrogen (77° K) to that of liquid oxygen (90° K), whereas the hydroxyl radicals would remain trapped in the water–hydrogen peroxide mixture up to 190° K.

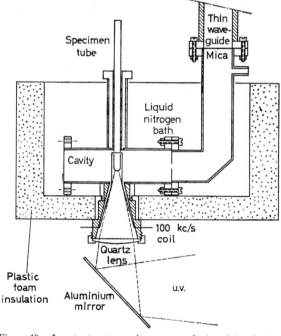

Figure 49. Low temperature cavity resonator for ' u.v.' irradiation ' in situ '. Reflection cavity for use with microwave bridge

These initial experiments demonstrated that electron resonance methods should provide a powerful new tool in such radical studies as these. It is not only possible to measure the radical concentrations quantitatively and study their mobility, but also to identify them and follow any radical reactions from the hyperfine structure of the spectra and any changes in it. In the initial experiments the samples were irradiated outside the apparatus by focussing a u.v. beam through a transparent quartz dewar onto the specimen which was cooled in liquid nitrogen. The specimen tube was then quickly removed and inserted into a precooled H_{012} cavity of the type shown in *Figure 15*, and it was estimated that the temperature did not rise by more than a few degrees in this process. In some applications, however, it is much better if the samples can be irradiated *in situ*, in the cavity, and the build up of radical concentration followed in detail. A cavity designed for this purpose is shown in *Figure 49*, and it can be seen how the u.v. beam is focussed on to the sample from below, through a thin-walled metal cone in the liquid nitrogen bath. This cavity is designed for operation with high frequency modulation and the loop producing this is also shown.

6.2.3 *Secondary Radicals*

Further work[5] along these lines then showed that secondary radicals could often be formed when radicals produced by the initial photolysis reacted with the solvent molecules. Hydroxyl radicals, formed from hydrogen peroxide, are very reactive in this way, and if a small amount of hydrogen peroxide is added to a hydrocarbon glass, spectra from secondary radicals, formed by proton abstraction from the hydrocarbon, are often obtained. The mechanism of this reaction is illustrated in *Figure 50(a)*, where the solvent is isopropyl alcohol. It is seen that the OH radicals formed by the u.v. irradiation each abstract an α–hydrogen from an isopropanol molecule leaving two isopropanol radicals each with an unpaired electron. If this interacts equally with the six protons of the two methyl groups, as shown, a seven-line hyperfine pattern is to be expected with a binomial distribution in its intensities. The spectra actually observed in this case, at 110° K, is shown in *Figure 50(b)*, and it is seen to be a seven-line spectrum with the correct intensity distribution. It is a general feature of the alcohol radicals that no interaction is obtained with the proton of the hydroxyl group. This may be due to rapid rotation of the –OH radical or to exchange of these protons through the medium via hydrogen bonding.

Figure 50(b) is typical of all the spectra observed from irradiated solids and it is interesting to compare it with those obtained from solutions, as discussed in the last chapter. There are two features that are noticeably different, first the width of the individual lines

174

is much larger, secondly the overall splitting of the spectrum tends to be greater. The increased line width is always to be expected in spectra of radicals in the solid phase, because narrowing due to

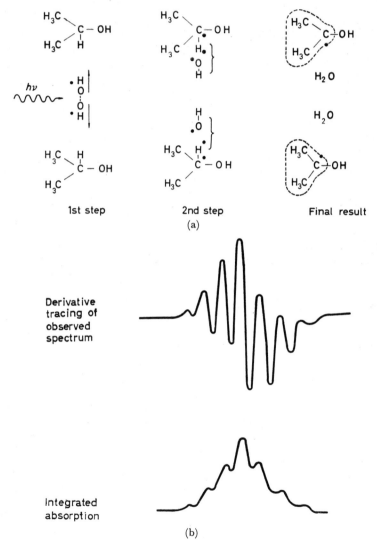

1st step   2nd step   Final result

(a)

Derivative tracing of observed spectrum

Integrated absorption

(b)

Figure 50. Secondary radical formation in isopropyl alcohol. (a) Mechanism of formation; (b) Derivative tracing of spectrum obtained, and its integrated absorption

the Brownian motion in a liquid is absent. The anisotropic hyperfine interaction is therefore not averaged out, and its summation over all possible orientations produces a line broadening of the isotropic hyperfine components. This effect has been discussed in detail in sections 4.3 and 4.4.1, and it was seen there that a broadening of about 25 gauss could be expected for normal C–H bond distances if all the protons were held rigidly in the solid. The width of the individual lines of *Figure 50(b)* is about 12 gauss and this is typical of most of the spectra observed in low temperature glasses. It would therefore appear that some type of partial motional narrowing is present in these cases. It has been suggested[6] that this is due to internal rotation of protons within the molecule, and that the broadening due to these is thus averaged to zero. On the other hand, the effect of protons in neighbouring molecules, locked in position in a solid matrix, still produces a finite average field and hence broadens the line. It should therefore be possible to determine under what conditions the internal motion of molecules is quenched, by studying the change of line width on cooling the sample. The expressions for the line width in the transition region, when the rate of molecular motion is comparable with the hyperfine splitting frequency have already been quoted and discussed in equations 4.27 and section 4.4.1.

The second main difference between spectra obtained by irradiation and those of aromatic radicals in solution is that the former usually have a larger extreme splitting. This is a somewhat accidental result since most of the irradiated compounds have been aliphatic molecules with groups for which a high degree of hyperconjugation is possible. This interaction can be considerably stronger than the configurational interaction of the aromatic ring protons, and hence wider overall splittings are obtained. Thus in the case of the isopropanol spectra of *Figure 50(b)*, the overall splitting is 120 gauss, and this is attributed to a hyperconjugative interaction with the six protons of the two methyl groups in the manner discussed in section 4.2.2. It is to be noted, however, that if a planar radical is formed, in which the hyperfine splitting is produced by configurational interaction only, then a much smaller overall splitting is obtained, similar to that of the aromatic ions. Such a case is provided by allyl alcohol from which secondary radicals can be formed after irradiation in the presence of a small amount of hydrogen peroxide[6]. The structure of the resulting radical and its electron resonance spectra are shown in *Figure 51*. It is seen that abstraction of an α–hydrogen leaves an unsaturated system with π bonds between the three carbon atoms. The attached

176

protons are therefore held coplanar with the carbon atoms by their sp^2 orbitals, and configurational interaction between the unpaired electron in the π–orbital system and the σ bonds of the C–H links is therefore possible. The splitting factor to be expected for a CH_2 group, and the accurate calculations of spin density for the allyl radical have been considered in section 4.2.1. It was seen that there is a spin density of $+0.622$ on each of the two end carbon atoms and of -0.241 on the central carbon. The splitting parameter, Q', was also estimated as 23 gauss, without allowing for the effect of the OH group, and hence the three equally-coupled protons on the end carbons should produce a quartet with overall spacing of $3 \times 23 \times 0.622 = 42$ gauss. The splitting obtained experimentally is 36 gauss, in good agreement with the predictions. The doublet splitting due to the central carbon atom would only be 5 gauss, and is not resolved in the observed spectrum.

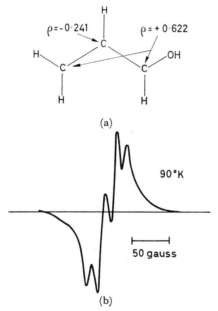

(a)

90°K

50 gauss

(b)

Figure 51. Spectrum of allyl alcohol radical. (a) Structural formula with predicted spin densities; (b) Derivative tracing of spectrum obtained (After Fujimoto and Ingram[6])

These examples of isopropanol and allyl alcohol therefore demonstrate that the hyperfine splittings in the spectra of radicals formed by irradiation can also be explained by the two effects of configurational interaction and hyperconjugation. A systematic survey of the radical species formed in various alcohols and polyhydric alcohols after hydrogen abstraction by photolysed hydrogen

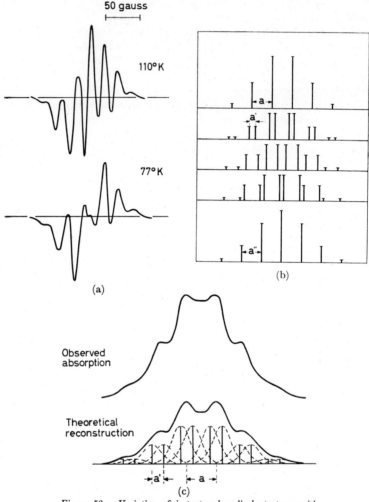

Figure 52. Variation of isopropanol radical spectrum with temperature. (a) Observed derivatives at 110° K and 77° K; (b) Change in line pattern as interaction with odd proton varies; (c) Reconstruction of absorption pattern at 77° K, and comparison with that obtained experimentally (After Fujimoto and Ingram[6])

peroxide has been carried out by GIBSON, INGRAM, SYMONS and TOWNSEND[5]. In most cases the hyperfine patterns could be explained on the assumption that an α–hydrogen had been abstracted by the primary hydroxyl radicals. In some cases two or more over-lapping spectra were obtained, however, showing that more than one radical species was present. In this connection it would appear that hydrogen bonding may play a major part in determining the geometry of the trapped radical, and thus the detailed hyperconju-gative interaction.

6.2.4 *Variation with Trapping Conditions*

Interesting information on the internal motion can also be obtained from the variation of the free radical spectra with tem-perature. This may be illustrated by the case of isopropyl alcohol, which was studied in some detail by FUJIMOTO and INGRAM[6]. Its spectra, after hydrogen abstraction by hydroxyl radicals, is shown in *Figure 52* as the temperature is lowered from 110° K to 77° K. At the higher temperature a seven-line spectrum is observed, indicating equal interaction with the six methyl protons, as shown in *Figure 50*. As the temperature is lowered and the glass becomes more viscous, the spectrum changes via a pattern with an extra central line at 90° K, to a symmetrical six-line spectrum, with a flat pit in the centre, at 77° K. It would therefore appear that some kind of quenching of the methyl group motion takes place at the lower temperature, so that one of the protons becomes less effective in hyperfine interaction. This type of inter-action might occur if the actual rotation of both methyl groups were quenched, and instead the protons in each group executed jumps between three equilibrium positions. One of the methyl groups would then be expected to have two of its protons in positions capable of hyperconjugative interaction with the carbon $2p_\pi$ orbital, and the third in the null plane, as in the structural diagram of section 4.2.2. The other methyl group would have its equilibrium positions rotated with respect to this, however, to avoid close proximity with the protons of the first group. All three of these protons could therefore interact with the carbon $2p_\pi$ orbital. This is probably too simple a picture in the intermediate tem-perature range, but might explain the main interactions at the higher and lower temperatures. If one proton becomes less effective under such a quenching process, the spectrum will consist of six lines, produced by the five equally coupled protons, each split into two by any residual interaction with the odd proton. The resulting spacing of the lines as the interaction with the odd proton, a', changes

179

from zero to a value equal to that of the other protons is shown in *Figure 52(b)*. The pattern obtained at 77° K can then be explained in detail if a' is taken as 9 gauss, and a, the splitting produced by the other five protons, is taken as 22 gauss. The reconstruction of the absorption spectrum and its comparison with that observed experimentally is shown in *Figure 52(c)*. It is very interesting to note in this connection that the splitting between the seven-line spectra at 110° K is given by $a''=20$ gauss, and that the relation

$$5a+a'=6a'' \qquad \qquad \dots (6.1)$$

is accurately obeyed. This implies that the amount of unpaired spin density which penetrates into the two methyl groups is conserved throughout the process. An alternative explanation of the six-line spectrum is that the rotation of both methyl groups is quenched giving an interaction with only four protons; but at the same time the rotation or exchange of the hydroxyl protons is also quenched. This proton would then also couple with the unpaired electron to give a total of five interacting protons and hence a six-line spectrum.

Variation of the spectra with temperature can also be used to differentiate between overlapping patterns when more than one radical is present. This may be illustrated by the case of methanol. The derivative of the spectra obtained from the methanol radical at 90° K, together with its integrated absorption, is shown in *Figure 53(a)*. It can be seen clearly from this that the total pattern is produced by overlapping three and five line spectra. The three-line spectrum is attributed to the $\dot{C}H_2 OH$ radical, the separation of 15 ± 2 gauss being that expected from configurational interaction in the CH_2 system (see section 4.2.1). It is suggested[6] that the five-line spectrum may be due to the formation of a $-\dot{C}H_2---H_2\dot{C}-$ biradical at certain points within the glass matrix, so that four equivalent protons contribute to the hyperfine pattern. This hypothesis is strengthened by the fact that the five-line spectrum increases in intensity as the temperature is raised. The derivative curve obtained from the methanol radical at 110° K is shown in *Figure 53(b)*, and it is seen that the extreme components in the wings have grown in size. The intensity of this five-line spectrum, is also increased by prolonged irradiation, as seen in the allyl spectra shown in *Figure 53(c)*. It is apparent that a five-line spectrum has now been formed as well as the original quartet. Hence higher temperatures, which will help reorientation, and prolonged irradiation, which will increase the probability of radical proximity, both produce a five line spectrum. This suggests that a biradical formed from two $-\dot{C}H_2$ groups may be a real possibility. There is

180

evidence for a similar superimposed five line spectrum in a large number of other spectra, and if the above hypothesis is correct, it would appear that transient biradical formation may be associated with the general decay mechanism of free radicals in a solid phase.

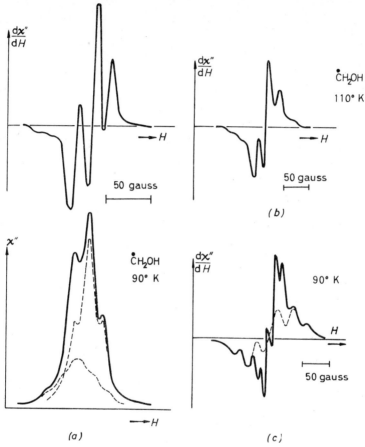

Figure 53. Occurrence of five-line spectra, thought to be due to $-\dot{C}H_2-H_2\dot{C}-$ biradicals. (a) Derivative tracing from methanol radical at 90° K, with its integrated absorption; (b) Methanol radical at 110° K; (c) Tracing of allyl radical spectrum after prolonged irradiation (After Fujimoto and Ingram[6])

6.3   X AND γ-RAY IRRADIATION. I. ORGANIC COMPOUNDS

The first electron resonance spectra observed from X-irradiated organic compounds were those obtained by SCHNEIDER, DAY and STEIN[8] from polymethylmethacrylate.   They irradiated the solid plastic with 200 kV X-rays to dosages of 10^5 to 10^7 roentgens, and produced quite a large electron resonance spectrum with a complex hyperfine pattern. The spectrum consists of five-line and four-line overlapping patterns, and its detailed interpretation has been the source of considerable speculation. In later work SCHNEIDER[7] suggested that it might be due to biradical formation at opposing ends of a broken polymer chain, but the appearance of an identical spectra during polymerization[9] does not support this view. It would seem, in fact, that the spectrum arises from two different mechanisms of interaction within the $R–CH_2–\dot{C}(CH_3)$ $(COOCH_3)$ radical[9,10]. Since most of the work on this spectrum has been performed during polymerization studies its detailed consideration is left to the next chapter. The derivative tracing together with the absorption curve are shown in *Figure 56* and analysed in detail in section 7.3.

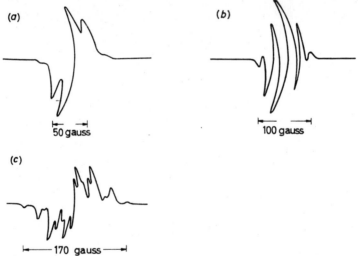

Figure 54. Spectra obtained from X-irradiated amino acids. (a) *Glycine;* (b) *Alanine;* (c) *Valine (After Gordy, Ard and Shields[11])*

In contrast to the first spectrum obtained, most of the others that are produced by X- or γ-ray irradiation possess hyperfine structure that can be readily interpreted in terms of the expected radical species. GORDY, ARD and SHIELDS[11,12] started a systematic investigation of the radicals formed when substances of biological interest were X-irradiated, and in most cases the hyperfine patterns showed that simple radical forms were produced as a net result of the radiation damage. The spectra that they obtained from various irradiated amino-acids are shown in *Figure 54*. These were observed after room-temperature irradiation with 50 kV X-rays. The triplet shown in *Figure 54(a)* is that obtained from glycine, CH_2 (NH_2) COOH, and it is apparent that the main hyperfine interaction is with two protons, and thus from a $\dot{C}H_2$ radical species. When X-rays or other high energy radiation is employed there are a large number of different ways in which the molecules may be broken up and various radicals formed. Thus it was suggested[11] that the primary act of irradiation might be to eject an electron and ionize the molecule which would then dissociate into small stable groups, such as NH_3 CO_2 and $(CH_2)^+$ in the case of glycine. The $(CH_2)^+$ radical might then stabilize itself by attachment to an adjacent $R\text{--}O^-$ ion. The disadvantage of X- and γ-irradiation studies is that sufficient energy is available to cause a variety of different breakdown processes, and it is usually impossible to differentiate between them. It is a striking feature of these studies, however, that simple radical species, containing relatively few interacting protons, are nearly always produced as the final product of the breakdown.

This fact is further illustrated in *Figure 54(b)* and *(c)*. The former is the spectrum obtained from irradiated alanine, $CH_3 . CH$ $(NH_2) . COOH$, and consists of a symmetrical five-line pattern. This therefore indicates that the unpaired electron is interacting with four equivalent protons, and it was postulated[11] that these were in a $(H_2 C\text{--}CH_2)^+$ group, the spin density reaching the proton $1s$ orbits by hyperconjugation. The spectrum obtained from irradiated valine, $CH_3 . CH(CH_3) . CH(NH_2) . COOH$, as shown in *Figure 54(c)*, consists of overlapping five and seven-line patterns, and the latter was postulated as arising from a $(H_3 C\text{--}CH_3)^+$ radical group. The overall splitting of this septet is somewhat larger than that obtained from the u.v. irradiated isopropyl alcohol[5], and this may well be due to partial electron-withdrawal by the OH group of the latter.

It can be seen, however, that the same basic theory of hyperfine interaction and hyperconjugation will account for these spectra in detail, as it did for the u.v. irradiated spectra, and those of the stable aromatic ions. The line widths are also of the same order as those obtained from the frozen u.v. irradiated glasses, showing that partial motional narrowing is also present in these systems to average out the dipolar interaction from the closest protons. The small shoulders in the wings of the irradiated glycine of *Figure 54(a)* are of interest in this connection. These may be due to the remnant dipolar interaction which might be resolved in this simple case.

Gordy, Ard and Shields[11,12] also studied a large number of other X-irradiated amino, carboxylic and hydroxy acids, and a summary of their results is given in *Table 6.1*. Most of these spectra could be explained in terms of overlapping patterns of n equally coupled protons, with n varying from 2 to 6. In some cases, however, broad symmetrical doublets were observed and these were thought to be due to dipole-dipole interaction between the unpaired electron localised on an oxygen or sulphur atom and a single proton held close to it for steric reasons. They also studied the effect of change of temperature on some of the carboxylic acid spectra. An interesting example of this is given by irradiated acetic acid. The spectra obtained from this after initial irradiation and observation at $77°$ K is shown in *Figure 55(a)*. It consists of a doublet with

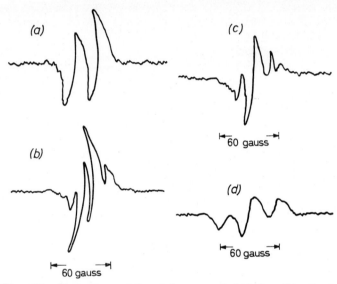

Figure 55. Temperature variation of the spectra obtained from X-irradiated acetic acid. (a) After initial irradiation and observation at $77°$ K; (b) After slight warming; (c) After warming to $190°$ K; (d) After recooling to $77°$ K (After Gordy, Ard and Shields[12])

25 gauss separation and lines of relatively broad width. The hydrogen bonding will probably hold the molecules fairly rigid at $77°$ K, and it is postulated[12] that the doublet arises from a direct dipole-dipole interaction between an electron localized on the hydroxyl oxygen, and the hydroxyl proton. On warming slightly a complete transformation of the spectrum occurs, as shown in *Figure 55(b)*, and a symmetrical quartet with narrow lines is obtained. This indicates equal coupling with three protons and is

almost certainly due to the methyl group, and the radical is postulated[12] as $CH_3 CO_2$. The narrow lines show that motional narrowing is now taking place and the dipolar interaction observed at the lower temperature is therefore averaged out. On further warming to 190° K the quartet changes to a triplet, as shown in *Figure 55(c)*. It was thought that this may be due to the formation of $(CH_2 CO)^+$, by a movement of the $CH_3 CO_2$ back to the ionized proton to form $(H_2 CCO)^+$ and water. If this sample at 190° K is then re-cooled to 77° K the triplet is found to double its spacing and broaden its lines as shown in *Figure 55(d)*. This suggests that the radical has ceased to rotate and that the unpaired electron, which resides mainly on the CH_2 carbon, now interacts through dipole-dipole coupling with the two protons so that a splitting additional to the isotropic term is produced, with a corresponding broadening.

It will be appreciated that the above explanations were only tentative first attempts, and other possible group formation may also account for the observed splittings. It can be seen, however, that this technique of watching radical conversion or decay, on warming the trapping medium, has great potentialities. One striking result of the measurements is to show how sensitive the rotation of different groups within the molecule can be to the exact viscosity of the surrounding medium.

A somewhat similar effect is observed in irradiated formic acid, HCOOH. The spectrum when initially at 77° K also shows a doublet, as in acetic acid. On warming to 190° K and re-cooling to 77° K two new lines appear on either side of the original doublet. One curious feature of this spectrum is that the splitting of the initial doublet is only 12 gauss, under half of that of the acetic acid. If the unpaired electron were again localised on the hydroxyl oxygen a similar splitting to that obtained for the acetic acid would be expected. It therefore seems that the electron may be localized on the carbonyl oxygen in the case of formic acid. The extra doublet produced after warming may be due to the shift of the unpaired spin density to the C–H bond with the possible formation of an HĊO radical[12].

Gordy, Ard and Shields[11] also studied the spectra obtained from biological material, such as proteins and bone tissue, after X-irradiation. By a comparison of these spectra with those obtained from the simpler acids, some idea of the breakdown processes occurring in natural products could be obtained. This work is summarized in more detail in the last chapter with the other biological and medical applications. Other workers have also studied spectra produced by X- or γ-irradiated organic compounds, and some of these are listed in *Table 6.1.* UEBERSFELD, COMBRISSON and ERB[13-15] have, in particular, made some quite detailed studies on γ-irradiated sugars, cellulose and similar material, and have recently obtained[16] an anisotropic spectrum from single crystals of glycine. A similar effect has also been observed by VAN ROGGEN *et al.*[49] with an X-irradiated single crystal of alanine. A considerable amount of work has also been undertaken by WHIFFEN *et al.*[10,17] on the spectra produced by γ-irradiation of polymers, and this is summarized in the next chapter with the other results of polymer study, including some of the interesting ' oxygen effects '.

Table 6.1. *Electron Resonance Spectra from X- or γ-Irradiated Organic Compounds*

Substance	Hyperfine pattern and overall splitting		Reference
Methyl alcohol HO—C—H (with H above and below)	Three lines	30 gauss	50
Ethyl alcohol HO—C—C—H (with H above and below)	Five lines	95 gauss	50
Acetamide C—C—H (with O, NH₂, H)	Three lines	45 gauss	50
Propionamide C—C—C—H (with O, NH₂, H)	Five lines	98 gauss	50
Acetanilide C—C—H (with O, H, N-phenyl)	Three lines	50 gauss	50

Substance	Hyperfine pattern and overall splitting		Reference
Sodium methoxide			

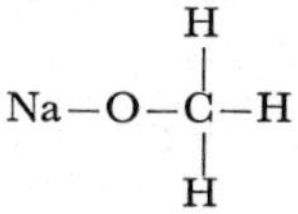

$$Na-O-\overset{\displaystyle H}{\underset{\displaystyle H}{C}}-H$$

| | Three lines | 30 gauss | 50 |

Mercury dimethyl

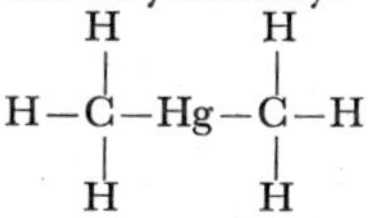

$$H-\overset{\displaystyle H}{\underset{\displaystyle H}{C}}-Hg-\overset{\displaystyle H}{\underset{\displaystyle H}{C}}-H$$

| | Five lines | 100 gauss | 51 |

Mercury diethyl

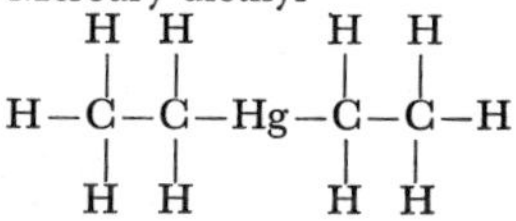

$$H-\overset{\displaystyle H}{\underset{\displaystyle H}{C}}-\overset{\displaystyle H}{\underset{\displaystyle H}{C}}-Hg-\overset{\displaystyle H}{\underset{\displaystyle H}{C}}-\overset{\displaystyle H}{\underset{\displaystyle H}{C}}-H$$

| | Six lines | 130 gauss | 51 |

Glycine

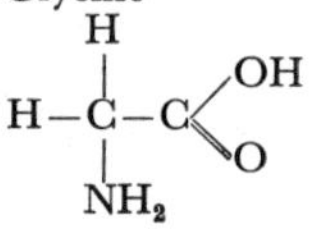

$$H-\overset{\displaystyle H}{\underset{\displaystyle NH_2}{C}}-C\overset{\displaystyle OH}{\underset{\displaystyle O}{\diagup}}$$

| | Three lines | 40 gauss | 11, 16, 18 |

Alanine

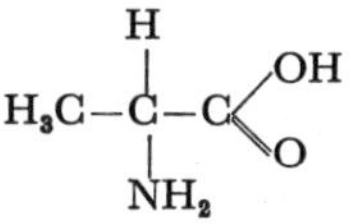

$$H_3C-\overset{\displaystyle H}{\underset{\displaystyle NH_2}{C}}-C\overset{\displaystyle OH}{\underset{\displaystyle O}{\diagup}}$$

| | Five lines | 90 gauss | 11 |

Valine

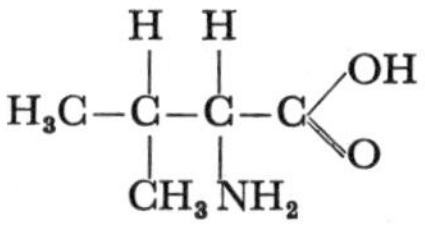

$$H_3C-\overset{\displaystyle H}{\underset{\displaystyle CH_3}{C}}-\overset{\displaystyle H}{\underset{\displaystyle NH_2}{C}}-C\overset{\displaystyle OH}{\underset{\displaystyle O}{\diagup}}$$

| | Seven lines plus five or seven lines | 170 gauss | 11 |

Leucine
$(CH_3)_2$ $CH.CH_2.CH(NH_2).COOH$

| | Two sets of five or more lines | 75 gauss | 11 |

Isoleucine
$CH_3.CH_2.CH(CH_3).CH(NH_2).COOH$ No resolvable structure

| | | | 11 |

Cystine
$(-S.CH_2.CH(NH_2).COOH)_2$

| | Asymmetric structure with four components | | 11 |

Glycolic acid

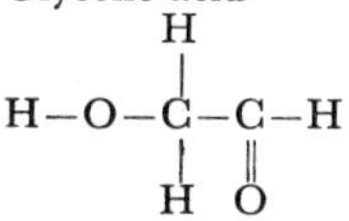

$$H-O-\overset{\displaystyle H}{\underset{\displaystyle H}{C}}-\overset{\displaystyle }{\underset{\displaystyle O}{C}}-H$$

| | Doublet | 20 gauss | 11 |

187

RADICALS PRODUCED BY IRRADIATION

Substance	Hyperfine pattern and overall splitting		Reference

Glycocyanine

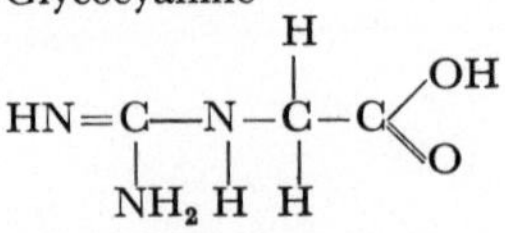

$$HN=C-N-C-C\begin{smallmatrix}OH\\O\end{smallmatrix}$$
with NH₂ H H below

Doublet — 15 gauss — 11

Polymethylmethacrylate

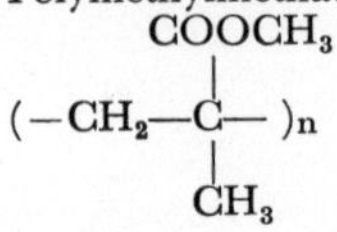

$COOCH_3$
$(-CH_2-C-)n$
CH_3

Five lines plus four lines — 100 gauss — 7, 8, 10

Polymethacrylic acid

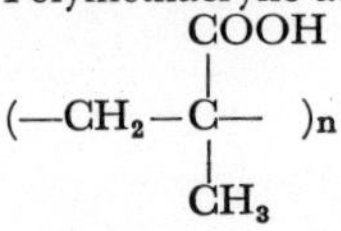

$COOH$
$(-CH_2-C-)n$
CH_3

Five lines plus four lines — 100 gauss — 10

Polyethylmethacrylate

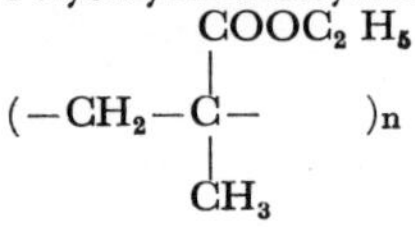

$COOC_2H_5$
$(-CH_2-C-)n$
CH_3

Five lines plus four lines — 100 gauss — 10

Polymethyl-chloro-acrylate

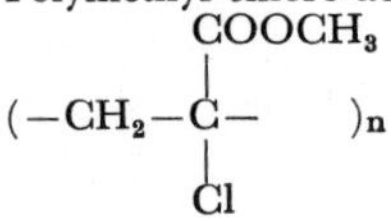

$COOCH_3$
$(-CH_2-C-)n$
Cl

Five lines — 40 gauss — 17

Polyacrylic acid

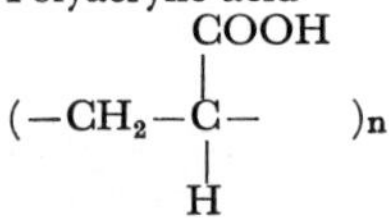

$COOH$
$(-CH_2-C-)n$
H

Three lines plus two — 50 gauss — 17

Polyvinyl alcohol

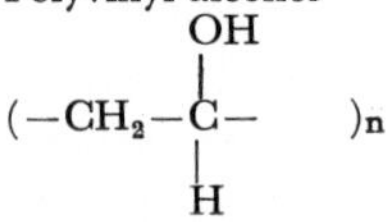

OH
$(-CH_2-C-)n$
H

Three lines — 70 gauss — 17

Polyvinyl acetate (hydrolysed)

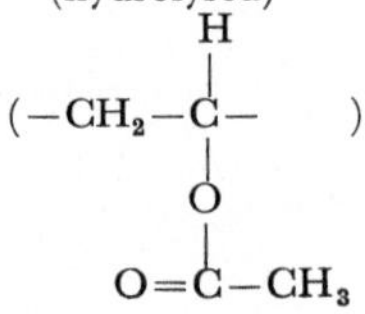

H
$(-CH_2-C-)n$
O
$O=C-CH_3$

Three lines — 60 gauss — 17

Substance	Hyperfine pattern and overall splitting		Reference
Polystrene $(-CH_2-\underset{\underset{\displaystyle \bigcirc}{|}}{\overset{\overset{\displaystyle H}{|}}{C}}-)_n$	Three lines	40 gauss	7, 17
Polythene $(-\underset{\underset{\displaystyle H}{|}}{\overset{\overset{\displaystyle H}{|}}{C}}-\underset{\underset{\displaystyle H}{|}}{\overset{\overset{\displaystyle H}{|}}{C}}-)_n$	Seven lines	90 gauss	17
Polyacrylonitrile $(-CH_2-\underset{\underset{\displaystyle H}{|}}{\overset{\overset{\displaystyle CN}{|}}{C}}-)_n$	Several lines about seven	110 gauss	17
Polytetrafluoroethylene $(-\underset{\underset{\displaystyle F}{|}}{\overset{\overset{\displaystyle F}{|}}{C}}-\underset{\underset{\displaystyle F}{|}}{\overset{\overset{\displaystyle F}{|}}{C}}-)_n$	Three lines	80 gauss	7
Nylon $(-CO(CH_2)_4.CO\ NH.(CH_2)_6.NH-)_n$	Four lines	80 gauss	17

If the hyperfine patterns summarized in *Table 6.1* are studied in detail, it will be seen that they are those normally expected from the number of protons included in the active group. There are some exceptions to this, however, such as the five-line spectra obtained from mercury dimethyl. One striking feature of the results is that protons of different groups all appear to be more or less equally coupled to the orbit of the unpaired electron.

6.4 X AND γ-RAY IRRADIATION. II. INORGANIC COMPOUNDS

Irradiated inorganic compounds were studied by electron resonance before the work on organic specimens was started. HUTCHISON[19] was the first to observe a resonance absorption, and this was obtained from lithium fluoride crystals irradiated for a few hours by a neutron flux of 10^{12} per cm^2 from an atomic pile. This work was followed by the X-irradiation studies of SCHNEIDER and ENGLAND[20], TINKHAM and KIP[21], and KIP, KITTEL, LEVY and PORTIS[22]. The paramagnetic entities produced by the irradiation in these alkali halide crystals can hardly be termed ' free radicals ' in the normal

189

sense of the word. They are better classified as 'damage centres' or 'defects' since they are associated with the structure of the crystal itself, rather than a broken bond in a particular molecule. A very brief summary of their study is nevertheless given below, in order to complete the picture of electron resonance studies on irradiated compounds.

The damage centres that are produced in single crystals by high energy irradiation can be grouped under three main headings (i) F-centres, (ii) V-centres, and (iii) interstitial atoms. An F-centre is produced when a negative ion is missing from the lattice and an unpaired electron is trapped in the coulomb field of the vacancy, and moves over the neighbouring ions. A V-centre is the opposite to an F-centre, i.e. when a positive ion is missing from the lattice, and an electron is thus also missing from one of the neighbouring ions. This leaves a 'hole' trapped in the vacancy, with its orbit distributed over the neighbouring ions. A V_2 centre is formed if two such V centres are produced side by side. The two electron holes then pair spins so that there is no resultant paramagnetism. A V_3 centre is a V_2 centre in which there is only one missing electron, or hole, between both vacancies, and this therefore has one unpaired spin like a V_1 centre, but with a more directional orbit. Interstitial atoms are produced when high-energy radiation, usually neutrons, knocks an atom from its normal lattice site into an interstitial position. This therefore leaves a vacant site as well as introducing an odd atom into the normal ionic binding of the lattice.

The above types of damage apply to pure single crystals, but if a hydride or deuteride is present in the lattice it is possible to form U-centres which are negative ion vacancies containing a hydrogen or deuterium negative ion. The unpaired electron is then mainly localized on the hydrogen and a large doublet hyperfine splitting is produced.

6.4.1   F and V Centres

A considerable amount of work was carried out on F and V centres before the double-resonance techniques became available. This work[20-23] showed that

(i) the g-value is below the free-spin value for F centres, which contain electrons, and above it for V centres, which contain holes;

(ii) the line width of F centres has a contribution from spin-lattice interaction above 90° K. Below this temperature the width is due to dipolar electron-spin broadening for concentrations of centres down to 10^{17} per c.c., but a residual width is then left due to unresolved hyperfine structure.

The fact that this width was due to hyperfine interaction and not to other forms of broadening was shown conclusively by PORTIS[24]. He studied the saturation behaviour of the signal and found no change in shape, thus indicating a mechanism of inhomogeneous broadening, as discussed in section 4.5.1.

These initial results on F centres were confirmed strikingly by the double resonance experiments of FEHER[25]. He was able to resolve out all the hyperfine splittings in the KCl F centre resonance by this technique, and measure the detailed coupling constants. This experiment has already been described in some detail in section 3.4, and the spectrum that he obtained is shown in *Figure 33*. Resolved hyperfine structure has also recently been obtained by LORD[26], who used normal techniques to study F centres formed in LiF and NaF. The fluorine nucleus has a larger magnetic moment than the chlorine and a spin of only one-half. A more resolved pattern is therefore to be expected and fifteen or more components were obtained in the spectrum. The theory for the molecular orbital interaction of the F centre's unpaired electron has been given by a number of authors[22,27-33], and is found to be in good agreement with the experimental results.

The first definite measurements on V centres were those of KANZIG[34]. These were produced by X-irradiation of KCl at $90°$ K and the spectrum was consistent with that from a Cl_2^- ion directed along a (110) axis. The g-value was found to be anisotropic, varying from $2 \cdot 0023$ to $2 \cdot 024$, and the hyperfine splitting from the two chlorine nuclei also varied markedly with angle. The vacancies were therefore identified as V_3 rather than V_1 centres, since the latter should have a cubic symmetry[35]. These results have been confirmed by recent measurements on X- and γ-irradiated thallium activated KCl[36], and the observed g-values can be accounted for very well by molecular orbital treatment[37].

6.4.2   *Interstitial Atoms and U centres*

Neutron irradiated diamond is a good example of a defect produced by interstitial atoms. The electron resonance spectrum of this has been studied by GRIFFITHS, OWEN and WARD[38], and a strong central line with a number of weak satellites is obtained. The central line has been attributed[39] to the single interstitial carbon atoms, while the satellites, which can be described by magnetic centres with a spin of one are attributed to double interstitial aggregates forming C_2 molecules. Trapping centres will usually be produced at the same time as interstitial atoms, as demonstrated by the measurements of WERTZ, AUZINS, WEEKS and SILSBEE[40] on neutron irradiated magnesium oxide. In this case electrons become trapped in the vacancies formed by the neutron irradiation, and a well resolved hyperfine splitting from the neighbouring Mg^{25} nuclei is obtained.

U-centres have been observed by DELBECQ, SMALLER and YUSTER[36] in irradiated mixed crystals of KCl–KH and KCl–KD. In this case the unpaired electron interacts mainly with the negative hydrogen or deuterium ion which is also trapped in the negative ion vacancy. As a result a large splitting is obtained, a doublet of 500 gauss separation for the hydrogen, and a triplet of 78 gauss separation for the deuterium. These values are very nearly the same as those for the hyperfine splitting of the free atoms and show that the unpaired electron must be highly localised on the hydrogen or deuterium. These spectra are very similar to those observed in γ-irradiated frozen acids, where free hydrogen atoms are postulated as trapped in the viscous glass, as discussed in more detail below.

6.4.3 γ-irradiation of ice and frozen acids

LIVINGSTON, ZELDES and TAYLOR[41] have carried out a series of investigations on the electron resonance spectra observed from γ-irradiated frozen acids. The three acids employed were H_2SO_4, $HClO_4$ and H_3PO_4, and each produced a doublet spectrum of about 500 gauss spacing as well as an absorption at $g=2\cdot0$. This doublet was attributed to free hydrogen atoms, which are formed by the γ-irradiation, and then become trapped in the frozen acid structure at 77° K. The fact that the hyperfine separation of the doublet was nearly the same for each of the different acids, and also very close to the separation in gaseous atomic hydrogen[42] confirmed the suggestion that the trapped species were hydrogen atoms, and not larger free radicals containing one interacting proton. This was further supported by the triplet structure which was obtained if the acid were diluted with D_2O before freezing and irradiation.

With sulphuric acid solutions the atomic hydrogen lines had the largest intensity for a 5 : 1 water to acid ratio, while this ratio was larger in $HClO_4$ and smaller in H_3PO_4. In every case, however, the lines were reduced in intensity for large dilutions and no free hydrogen was observed in irradiated ice itself. These measurements were made at 9,000 Mc/s and 25,000 Mc/s, whereas measurements were also made by SMALLER, MATHESON and YASAITIS[43,44] on γ-irradiated ice at 350 Mc/s. They obtained a doublet in this case, and also attributed it to atomic hydrogen, but the separation was only 30 gauss. In the same way, irradiated D_2O gave a triplet, but with a correspondingly reduced splitting. This reduction of the splitting from the free atom value was explained as due to the large polarization effect of the ice, but it is a little difficult to see why this should not also apply to the more dilute acid solutions studied by LIVINGSTON, ZELDES and TAYLOR[41].

There still seems to be some mystery about this problem, since, although the large splittings obtained from the acids were confirmed at 350 Mc/s[44], the small splitting obtained from the pure ice has not been confirmed in the microwave region[41]. On the other hand, MATHESON and SMALLER[44] were able to obtain distinctly different spectra from γ-irradiated H_2O_2 or D_2O_2 water mixtures, showing that the initial spectra were definitely not due to OH or HO_2 radicals.

Two other features of interest also emerged from the work on the frozen acids. One was that the hyperfine lines of the atomic hydrogen had weak satellites on either side at a separation corresponding to the proton resonance frequency in the same magnetic field[45]. These arise from simultaneous transitions of nearby proton spins at the same time as the electron transition. One essential condition for their appearance is that the hydrogen atom is held rigidly in position, and they can thus be used as a sensitive probe to indicate the onset of any diffusive motion. The second feature of interest was that detailed decay rates were measured at different temperatures[41]. These were shown to fit well into a second order kinetic process, and confirm that the main mechanism of atomic hydrogen removal is by recombination with itself. The spectra will disappear instantly, of course, when the frozen acids are melted.

6.5   OTHER IRRADIATED INORGANIC COMPOUNDS

Some measurements have also been made on irradiated nitrates. Thus ARD[46] found a resonance from X-irradiated sodium nitrate, and BLEANEY, HAYES and LLEWELLYN[47] also found spectra from irradiated lanthanum magnesium nitrate. In both cases a triplet hyperfine splitting was obtained, and attributed to an interaction with the nitrogen atom in NO_2 molecules which were thought to be formed as a result of the irradiation. It is noticeable in this respect that the observed hyperfine splittings of about 50 gauss are very similar to those obtained in the free gas[48]. The irradiation damage in the lanthanum magnesium nitrate was produced initially by small amounts of radioactive americium or plutonium ions embedded in the lattice, and spectra arising from this source must be carefully watched when radioactive elements are being studied.

A large number of gaseous free atoms or radicals have also been produced by u.v. irradiation and their electron resonance spectra studied. These results are summarized and discussed with the work on stabilized inorganic radicals in Chapter 8. It will be seen from the different sections in this chapter that the study of irradiation physics and chemistry by electron resonance techniques is now reaching into a large number of different fields, and is providing some very detailed information on various solid state and molecular interactions.

REFERENCES

[1] INGRAM, D. J. E., HODGSON, W. G., PARKER, C. A. and REES, W. T. *Nature, Lond.* 176 (1955) 1227

[2] BIJL, D. and ROSE-INNES, A. C. *Ibid.* 175 (1955) 82

[3] LEWIS, G. N. and LIPKIN, D. *J. Amer. chem. Soc.* 64 (1942) 2801

[4] NORMAN, I. and PORTER, G. *Proc. roy. Soc. A* 230 (1955) 399

[5] GIBSON, J. F., INGRAM, D. J. E., SYMONS, M. C. R. and TOWNSEND, M. G. *Trans. Faraday Soc.* 53 (1957) 914

[6] FUJIMOTO, M. and INGRAM, D. J. E. *Ibid.* 54 (1958) 1304

[7] SCHNEIDER, E. E. *Disc. Faraday Soc.* 19 (1955) 158

[8] — DAY, M. J. and STEIN, G. *Nature, Lond.* 168 (1951) 645

[9] INGRAM, D. J. E., SYMONS, M. C. R. and TOWNSEND, M. G. *Trans. Faraday Soc.* 54 (1958) 409

[10] ABRAHAM, R. J., MELVILLE, H. W., OVENALL, D. W. and WHIFFEN, D. H. *Ibid.* 54 (1958) 1133

[11] GORDY, W., ARD, W. B. and SHIELDS, H. *Proc. Nat. acad. Sci.* 41 (1955) 983

[12] GORDY, W., ARD, W. B. and SHIELDS, H. *Ibid.* 41 (1955) 996

[13] UEBERSFELD, J. *C. R. Acad. Sci., Paris,* 239 (1954) 240

[14] — *Ann. Phys.* 1 (1956) 395

[15] COMBRISSON, J. and UEBERSFELD, J. *Disc. Faraday Soc.* 19 (1955) 181

[16] UEBERSFELD, J. and ERB, E. *C. R. Acad. Sci., Paris,* 242 (1956) 478

[17] ABRAHAM, R. J. and WHIFFEN, D. H. *Trans. Faraday Soc.* 54 (1958)

[18] EHRENBERG, A., EHRENBERG, L. and ZIMMER, K. G. *Acta. chem. Scand.* 11 (1957) 199

[19] HUTCHISON, C. A. *Phys. Rev.* 75 (1949) 1769

[20] SCHNEIDER, E. E. and ENGLAND, T. S. *Physica,* 17 (1951) 221

[21] TINKHAM, M. and KIP, A. F. *Phys. Rev.* 83 (1951) 657

[22] KIP, A. F., KITTEL, C., LEVY, R. A. and PORTIS, A. M. *Ibid.* 91 (1953) 1066

[23] HUTCHISON, C. A. and NOBLE, G. A. *Ibid.* 87 (1952) 1125

[24] PORTIS, A. M. *Ibid.* 91 (1953) 1071

[25] FEHER, G. *Ibid.* 105 (1957) 1122

[26] LORD, N. W. *Ibid.* 105 (1957) 756

[27] MUTO, T. *Prog. theor. Phys., Japan,* 4 (1949) 243

[28] INUI, T. and UEMURA, Y. *Ibid* 5 (1950) 252, 395

[29] KAHN, A. H. and KITTEL, C. *Phys. Rev.* 89 (1953) 315

[30] KITTEL, C. *Defects in Crystalline Solids.* Phys. Soc. 1955, p. 33

[31] DEXTER, D. L. *Phys. Rev.* 93 (1954) 244

[32] KRUMHANSL, J. A. *Ibid.* 93 (1954) 245

[33] SEITZ, F. *Rev. mod. Phys.* 26 (1954) 7

[34] KANZIG, W. *Phys. Rev.* 99 (1955) 1890

REFERENCES

[35] COHEN, M. H. *Ibid.* 101 (1956) 1432

[36] DELBECQ, C. J., SMALLER, B. and YUSTER, P. H. *Ibid.* 104 (1956) 599

[37] INUI, T., HARASAWA, S. and OBATA, Y. *J. phys. Soc., Japan,* 11 (1956) 612

[38] GRIFFITHS, J. H. E., OWEN, J. and WARD, I. M. *Defects in Crystalline Solids.* Phys. Soc. 1955, p. 81

[39] O'BRIEN, M. C. M. and PRYCE, M. H. L. *Ibid.* p. 88

[40] WERTZ, J. E., AUZINS, P., WEEKS, R. A. and SILSBEE, R. H. *Phys. Rev.* 107 (1957) 1535

[41] LIVINGSTON, R., ZELDES, H. and TAYLOR, E. H. *Disc. Faraday Soc.* 19 (1955) 166

[42] NAFE, J. E. and NELSON, E. B. *Phys. Rev.* 73 (1948) 718

[43] SMALLER, B., MATHESON, M. S. and YASAITIS, E. L. *Ibid.* 94 (1954) 202

[44] MATHESON, M. S. and SMALLER, B. *J. chem. Phys.* 23 (1955) 521

[45] ZELDES, H. and LIVINGSTON, R. *Phys. Rev.* 96 (1954) 1702

[46] ARD, W. B. *J. chem. Phys.* 23 (1955) 1967

[47] BLEANEY, B., HAYES, W. and LLEWELLYN, P. M. *Nature, Lond.* 179 (1957) 140

[48] CASTLE, J. G. and BERINGER, R. *Phys. Rev.* 80 (1950) 114

[49] VAN ROGGEN, A., VAN ROGGEN, L. and GORDY, W. *Bull. Amer. phys. Soc.* 1 (1956) 266

[50] LUCK, C. F. and GORDY, W. *J. Amer. chem. Soc.* 78 (1956) 3240

[51] GORDY, W. and MCCORMICK, C. G. *Ibid.* 78 (1956) 3243

RADICALS FORMED BY POLYMERIZATION AND PYROLYSIS

7.1 Introduction

THE electron resonance spectra obtained during polymerization processes, and during the charring or coalification of organic matter, are summarized in this chapter. These two systems have one notable feature in common in that the free radicals are 'self-trapping', and relatively large concentrations of stabilized radicals are built up as the process continues. It is therefore not necessary to employ artificial trapping techniques as in the u.v. irradiation studies. The two systems also have other features in common, such as the decay of free radical signal on interaction with molecular oxygen, and the results obtained with these different types of high molecular weight organic radicals have therefore been brought together in this chapter.

Radical processes have been postulated as the mechanism of polymer growth for some considerable time, and in the ideal electron resonance experiment it would be possible to watch the growth and decay of dynamic concentrations of the free radicals as the polymerization proceeded. Unfortunately the magnitudes of these dynamic concentrations are only just within the range of the highest sensitivity spectrometers at the moment, and detailed measurements have yet to be made on these systems. The actual process of polymerization affords a very effective trapping mechanism for the active radical ends of the growing chains, however, and considerable information on the polymerization process can be obtained by studying this concentration of trapped radicals under different conditions. The trapping is a purely steric effect in this instance and takes place either when the polymer gels at a stage of partial polymerization, or when the polymer is insoluble in the monomer and precipitates out around the growing chains. In either case a solid saturated medium is formed around the active end of the chain, which is therefore prevented from interacting with more monomer, and remains trapped with its unpaired electron in a polymer cage.

The results obtained from these polymer 'gels' or precipitates are discussed first, and then a summary is given of the γ–irradiation studies, as these give additional information on the nature of the active ends. The different effects of molecular oxygen on the polymer radicals are also discussed in this connection. In the second half of the chapter the work on free radicals trapped in charred organic matter is outlined, together with the various hypotheses which have been put forward to account for the high concentration and stability of this type of radical. It would appear that the trapping mechanism here is again one of physical size, but size associated with a system of condensed carbon rings and hence with large resonance energy.

7.2 INITIAL WORK ON POLYMERIZATION

The first study of polymerization by electron resonance was by FRAENKEL, HIRSHON and WALLING[1] who studied a series of cross-linked vinyl polymers in a gelled state. The monomer and initiator were heated at 60–100° C in a sealed glass tube until gelation occurred, and the electron resonance of the sample was then observed at room temperature. The first system studied was glycol dimethacrylate, and they found that identical spectra were produced by a large number of different initiators such as, benzoyl peroxide, azobisisobutyronitrile, t-butyl-perbenzoate, and di-t-butyl peroxide. The same spectrum was also obtained if methylmethacrylate were used, and the spectra was in fact identical with that observed by SCHNEIDER, DAY and STEIN[2] on X–irradiation of polymethylmethacrylate. It is shown in *Figure 56* and its interpretation is discussed in detail in the next section. Spectra were also obtained[1] from gels of vinylmethacrylate, divinyl benzene and glycol diacrylate, and these initial measurements demonstrated that electron resonance can give some very useful additional information on the nature of the active polymer ends.

Experiments on polymers which are insoluble in the monomer and precipitate out during the polymerization process were then undertaken by BAMFORD, JENKINS, INGRAM and SYMONS[3]. In this case, too, it was shown that the active radical ends became trapped in the ' dead polymer cage ', and quite high concentrations of radicals were observed in polyacrylonitrile. The monomer was polymerized photochemically in this case, with di-t-butyl peroxide as a catalyst, and after 20 per cent conversion the unchanged monomer and catalyst were distilled off in vacuum and the tube sealed. Radical concentrations of up to 10^{17} per gramme were observed in polyacrylonitrile prepared in this way.

7.3 POLYMETHYLMETHACRYLATE

Following these initial polymerization studies several series of systematic experiments have been performed, and the electron

197

resonance spectra of quite a large number of different radicals have been observed. This work has so far been carried out mainly by INGRAM, SYMONS and TOWNSEND[4] and by MELVILLE, WHIFFEN et al.[5, 6]. One of the first spectra to be studied in detail was that of polymethylmethacrylate, since the initial work had provided no simple explanation of its complete hyperfine pattern. The structural formula of polymethylmethacrylate is shown in *Figure 56(a)*, and the derivative tracing and integrated absorption curve of its spectrum in *Figure 56(b)*. It can be seen that the hyperfine

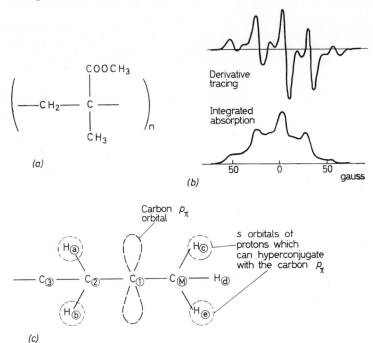

Figure 56. Spectrum of the polymethylmethacrylate radical. (a) *Structural formula;* (b) *Derivative tracing of hyperfine structure, and its integrated absorption;* (c) *Hyperconjugative interaction within the radical*

pattern consists of overlapping five and four-line spectra, and the width of the individual lines is about eight gauss. The five-line spectrum has components spaced at 0, $\pm$ 26, and $\pm$ 52 gauss from the centre, with relative intensities of $1:4:6:4:1$ as would be expected for equal interaction with four protons. The four-line spectrum, which is somewhat weaker in intensity, has components spaced at $\pm$ 13 and $\pm$ 39 gauss from the centre with relative intensities of $1:3:3:1$, as expected for interaction with three equivalent protons. It therefore appeared that two radical species must be present, one with four equivalent protons and one with three equivalent protons. The striking feature of this spectrum, however, is that an identical pattern is obtained whether the radicals are produced by γ–ray, X-ray or u.v. irradiation or by occlusion in gelled or precipitated polymers. It was thus very difficult to see how two different radical species could always be produced in the same relative proportions by such different methods. On the basis of the X-ray measurements, SCHNEIDER[7] suggested that the spectrum could be explained by a biradical interaction between two unpaired electrons on the opposing ends of a severed polymer chain. The fact that an identical spectrum is also obtained from the radicals formed during polymerization[4,5], in which no biradicals are likely to be formed, is rather strong evidence against this hypothesis, however.

ABRAHAM, MELVILLE, OVENALL and WHIFFEN[5] then performed a series of experiments on γ–irradiated polymers derived from polymethylmethacrylate by replacing different groups. They found that identical spectra were obtained from irradiated polymethacrylic acid (in which the $-COOCH_3$ is replaced by $-COOH$) and from polyethylmethacrylate ($-COOCH_3$ replaced by $-COOC_2H_5$), as from the polymethylmethacrylate itself. On the other hand, if the main chain methyl group is replaced, as in polymethyl-α-chloroacrylate or polyacrylic acid, quite different spectra are obtained. These experiments showed conclusively that the mixed four and five-line pattern must therefore originate from a radical

$$-CH_2-\overset{\displaystyle COOR}{\underset{\displaystyle CH_3}{\overset{\displaystyle |}{\underset{\displaystyle |}{C}}}}\cdot$$

of the form and that the orbit of the unpaired electron does not embrace R, which can be H, CH_3 or C_2H_5. This work was further evidence against an explanation of the spectrum in terms of two different radicals, and it would appear that the whole spectrum must therefore be due to this one radical. There is therefore a maximum of five protons with which the unpaired electron can interact, and in order to explain the existence of both a five and a four line spectrum it must be assumed that the electron sometimes interacts with four of these, and sometimes with three. This assumption at first seems very unlikely, but it can be justified quite readily if the possible mechanisms of hyperconjugative interaction are considered in detail.

Two slightly different explanations[4,5] were put forward, more or less simultaneously, to account for the observed spectrum by such hyperconjugative interaction and both can be summarized by reference to *Figure 56(c)*. In this diagram the chain of carbon atoms is represented as C_1 C_2 C_3 $--$, the methyl carbon attached to the main chain is represented by C_M, and the $-COOR$ group has been omitted since it has no interaction with the unpaired electron's orbital. To a first approximation the unpaired electron will be located in the p_π orbital of C_1 as shown by the dotted lobes. It has already been shown in section 4.2.2 that it is possible for this electron to interact with the protons H_c, H_d, and H_e of the methyl group, by hyperconjugation. If this methyl group is rotating with a frequency faster than that of the hyperfine splitting there will be an equal interaction with all three protons. If, on the other hand, the methyl group remains rigid the orbitals will be as shown in the structural diagram of section 4.2.2, and hyperconjugative inter-action will only take place with the two protons H_c and H_e. In the same way hyperconjugation will also take place between the electron in the p_π orbital of C_1 and the two protons H_a and H_b of the CH_2 group. These cannot rotate, however, since the position of the third proton is now taken by C_3 and the rest of the polymer chain. This CH_2 group need not be orientated as shown, however. It is possible for the positions of either H_a or H_b to be interchanged with that of C_3. In this case interaction would only occur with the remaining proton. It is also possible for the C_3 H_a H_b group to be rotated by $30°$ about the C_1 C_2 axis so that only one of the three atoms lies in the plane suitable for hyperconjugative inter-action. If this atom is C_3 there will be no extra proton hyperfine interaction from this group, whereas if it is a hydrogen there will be one extra proton coupled to the interacting system.

Abraham, Melville, Ovenall and Whiffen[5] suggested that the observed four and five-line spectrum could be accounted for if the methyl group were fixed, giving an interaction with two protons, and the CH_2 group was also fixed, but sometimes with the atoms distributed as shown in *Figure 56(c)*, and sometimes with C_3 inter-changed by H_a or H_b. There would then be an extra coupling with either two or one proton and hence a total interaction with either four or three protons, as required to explain the hyperfine pattern. Ingram, Symons and Townsend[4] suggested, on the other hand, that the methyl group was rotating all the time, giving an interaction with three protons, and that the orientation of the C_3 H_a H_b group had been rotated by the $30°$ so that sometimes C_3 and sometimes H_a or H_b was in the plane of hyperconjugative interaction. This again results in two different structures of the radical with either three or four protons coupled to the unpaired electron. It can be seen that these two hypotheses are very similar in principle, and the only practical difference is that the former assumes that the methyl group is held rigidly in the polymer while the latter assumes that it is rotating. It is interesting, in this connection, that the radical formed by hydroxyl attack on the monomer at liquid nitrogen temperatures has a six-line spectrum[4]. In this case the CH_2 group no longer has the rest of the polymer chain attached, and might be expected to rotate as well. Both H_a and H_b will then interact equally with the p_π orbital at the same time as the methyl hydrogens, and a six-line spectrum from five equally coupled protons is therefore to be expected.

The analysis and explanation of the polymethylmethacrylate hyperfine pattern is a very good example of the detailed information that can be obtained on radical structure if the precise mechanism of interaction is carefully analysed. Most hyperfine patterns that have been obtained so far have been easier to explain than this case, but, on the other hand, have not given so much detailed information on the structure and orientation of the groups within the radical.

7.4   Effects of Adsorbed Oxygen

It was noticed in the initial electron resonance studies on polymers[1, 3] that the signal decayed as soon as oxygen was admitted to the specimen, the decay time being determined by the diffusion of the oxygen through the sample. This decay was postulated as due to the initial formation of $\equiv$ C–O–O$^\bullet$ radicals with subsequent change to $HO_2^\bullet$ or similar species which could react to form non-radical products. Ard, Shields and Gordy[8] then found that for some X-irradiated polymers, the radicals formed by interaction with the molecular oxygen became stabilized in the system and remained permanently trapped. An example of this is irradiated tetrafluoro-ethylene $(-CF_2-CF_2-)_n$. The spectrum of this after X-irradiation and before ageing in air is shown in *Figure 57(a)* and consists of eight symmetrical components with an additional resonance superimposed near the centre of the first group. The eight-line spectrum is assumed to be due[8] to an interaction with seven of the fluorine nuclei attached to the chain (F^{19} has a spin $I = \frac{1}{2}$). Thus the fission of the polymer chain by the high energy irradiation may cause the severed ends to coil back and contract, bringing the orbital of the end carbon atom close to several fluorine nuclei. This recoil process will also leave a relatively deep vacancy in the structure and it is possible that an oxygen molecule will become trapped in this and held permanently in position.

The central resonance of *Figure 57(a)* is in fact attributed[8] to the effect of such absorbed oxygen, since on ageing in oxygen the eight-line spectrum vanishes while the central resonance grows in size. The spectrum after ageing in air for two weeks is shown in *Figure 57(b)*. If the irradiated polymers were aged in nitrogen or in vacuum no change from the original eight-line spectrum was observed. It is therefore assumed that the oxygen molecules diffuse into the solid polymer, become trapped in the vacancies caused by the irradiation, and interact with the unpaired electron of the polymer radical to localize this close to the oxygen molecule. The

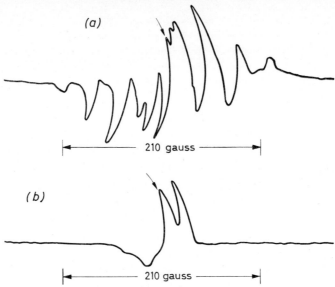

Figure 57. Spectra from X-irradiated Teflon. (a) *Immediately after irradiation;* (b) *After ageing in oxygen. The arrow indicates the position of the free-spin g-value (After Ard, Shields and Gordy[8])*

resultant complex can then be regarded as having a three electron bond between the two oxygen atoms, i.e. $-F_2C-O \ddot{-} O$. It is noticeable in this connection that the central resonance can be resolved into a triplet as seen in *Figure 57(b)*, and that the splitting of this triplet is dependent on the direction of the magnetic field. This suggests that the splitting may be partially due to an anisotropic electronic interaction produced by strong localization of the unpaired electron on the oxygen atoms.

202

It would therefore appear that the adsorption of molecular oxygen can have two opposite effects on an electron resonance spectra. In the first type of interaction, which is the more common, the oxygen forms a transient R–O–O˙ radical with the initial group, and this then breaks up into HO$_2$˙ or similar radicals which react to give non-radical products. The net effect of this on the observed spectrum is a gradual decay to zero. In the second type of inter-action the oxygen also forms an R–O–O˙ radical but this then remains stabilized or trapped in the system. The unpaired electron is now strongly localized on the oxygen atoms and any hyperfine structure due to a large molecular orbit is thus removed, and its place taken by a relatively wide central line which may have asym-metry due to anisotropic g–values or electronic splittings.

Several examples of both of these types of interaction have been observed in different γ–irradiated polymers. Two cases of particular interest are low pressure polyethylene and poly-chlorotrifluoro-ethylene[9]. These show no electron resonance absorption after γ-irradiation in vacuum before admission of oxygen. On admission of oxygen a single asymmetric resonance is obtained, however, with a rate of growth that is diffusion controlled. This is attributed to the formation of stable R–O–O˙ radicals at the damage centres. It is noticeable that the same spectra are obtained directly after γ–irradia-tion in air. The complete absence of any spectra before oxygen admission is somewhat puzzling but may be due to strong interaction between biradicals on opposite ends of a severed bond. This could produce an anisotropic splitting and broaden the resonance beyond detection.

Another feature of interest in these polymers is that the oxygen effect is reversible if the polymer is held at about 80° C, and the spectra will disappear again if the polymer is re-evacuated at this temperature[9]. A similar reversible interaction with gaseous oxygen was first noticed in the case of the radicals trapped in low tempera-ture carbons and these effects are discussed in more detail in section 7.7.3.

7.5 VARIATION OF RADICAL CONCENTRATIONS DURING POLYMERIZATION

It has already been noticed that the steady state concentrations of free radicals taking part in a polymerization process are only just on the verge of possible detection (10^{-8} moles/litre). This concentration can be increased by employing the ' gel ' or ' pre-cipitation ' effect, however. The viscosity of the medium then becomes so large that the diffusion of the radical ends is greatly reduced and the rate of mutual chain termination thus falls rapidly. The viscosity can also be increased by the use of suitable cross-linking agents, as in the initial experiments of Fraenkel, Hirshon and Walling[1].

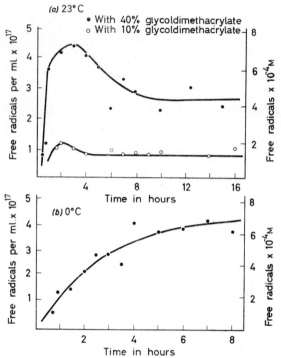

Figure 58. Variation of radical concentration with time, during copolymerization of methylmethacrylate and glycoldimethacrylate. (a) At 23° C. Top curve is for 40% glycoldimethacrylate. Bottom curve is for 10% glycoldimethacrylate; (b) At 0° C, with 40% glycoldimethacrylate (After Atherton, Melville and Whiffen[10])

The kinetics of such systems have recently been studied in more detail by ATHERTON, MELVILLE and WHIFFEN[10]. They investigated the variation of radical concentration during the copolymerization of methylmethacrylate and glycoldimethacrylate, and its dependence on the constitution of the mixture. The polymerization was initiated photochemically with azobisisobutyronitrile as a catalyst. The variation of radical concentration with time of irradiation at room temperature is shown in Figure 58(a) for two different polymer mixtures. The top curve is for a mixture containing 40 per cent glycoldimethacrylate, while the bottom is for that containing only 10 per cent of the copolymer. It is evident that a larger radical concentration is found with the higher percentage of copolymer, and thus higher cross-linking, as would be expected. The occurrence of a maximum in both curves after about an hour's irradiation is due to the natural decay of radicals at room temperature. This rate of decay is larger than the rate of formation immediately after the initial radicals have been produced. Its effect can be eliminated, however, if all the measurements are made on samples held at 0° C. The corresponding curve for the 40 per cent glycol dimethacrylate mixture held at 0° C is shown in *Figure 58(b)*. The rate of radical decay is now very small and a steady rise in the radical concentration is observed. It may be noted that the nature of the radical decay at room temperature suggests that the radical ends are initially held in the cross-linked network in such a manner that some pairs of ends can reach each other by diffusion[10]. If this is going to take place it does so during the first few hours after initial formation.

It is also possible to investigate the operation of inhibitors by kinetic studies of radical concentration. An example of this is the work of HARLE and THOMAS[39] on the inhibiting action of phenyl–α–naphthylamine on the oxidation of octadecene. There are several alternative proposals[40,41] for the actual mechanism of inhibition, and one of these involves an equilibrium concentration of a radical complex. Harle and Thomas[39] therefore followed the concentration of radicals formed during the oxidation of octadecene at 171° C when inhibited by phenyl–α–naphthyalmine at 0·0108 M initial concentration. Samples were withdrawn from the reaction at different time intervals and immediately frozen in liquid nitrogen, and the radical concentration then determined by an *X*–band electron resonance spectrometer. In this way the dynamic radical concentration could be plotted against time of interaction. It was found to have a peak after three and a quarter hours, and then to fall off rapidly in the following half-hour. The variations of inhibitor concentration and of oxygen consumption were also measured and the complete set of kinetic data were then compared with those predicted by the alternative theories of inhibitor interaction. The fact that the radicals were associated with the intermediates produced by the inhibitor was confirmed by their complete absence in the unoxidized sample, in the inhibitor itself, and in the octadecene oxidized in the absence of the inhibitor.

There is no doubt that these initial kinetic studies are only the beginning of the possible applications of electron resonance in these fields, but they serve as a good illustration of the kind of experiments that will be possible with increasing sensitivity in the future.

Table 7.1 Polymer Radicals

Monomer	Method of Radical formation	Radical Concentration	Hyperfine Pattern	Ref.
Acrylonitrile	Occlusion in ppted polymer	5.10^{-4} M	Single line	**3, 4**
	Trapped in low temp. glass	5.10^{-2} M	Unresolved line	**4**
	γ-irradiation	10^{-3} M	About 9 lines	**11**
Acrylic Acid or methyl ester	Occlusion in ppted polymer	10^{-5} M	Three lines	**12**
	γ-irradiated	10^{-3} M	Three lines	**11**
Divinylbenzene and Glycoldiacrylate	Occlusion in cross-linked gel	10^{-4} M	Three lines	**1**
Methacrylonitrile	Occlusion in ppted polymer	10^{-5} M	Overlapping 5 and 4 line patterns.	**4**
	Trapped in low temp. glass	5.10^{-2} M	6 lines	**4**
Methylmeth-acrylate	Occlusion in ppted polymer	10^{-5} M	Overlapping 5 and 4 line patterns	**4**
	Occlusion in gel and cross-linked system	10^{-4} M	Overlapping 5 and 4 line patterns	**1, 6**
	X-irradiation	10^{-4} M	Overlapping 5 and 4 line patterns	**7**
	γ-irradiation	10^{-3} M	Overlapping 5 and 4 line patterns	**11**
	Trapped in low temp. glass	10^{-2} M	6 lines	**4**
Vinyl bromide	Occlusion in ppted polymer	No absorption		**4**
	γ-irradiation	10^{-3} M	1 line	**11**
	Trapped in low temp. glass	10^{-2} M	Single broad line	**4**

7.6 OTHER POLYMER STUDIES

The electron resonance spectra of several other polymers have also been studied, and the radicals have been formed either by occlusion in gels or precipitates, or by irradiation. Some of the irradiation results have in fact already been summarized in *Table 6.1*. The results on polymer radicals produced by different means are also brought together for comparison in *Table 7.1*. It will be seen from this that the same hyperfine pattern is usually produced, whatever the method of radical formation, indicating that the same basic group is associated with the end of a chain, even after high energy irradiation. One notable exception to this rule is the case of polyacrylonitrile where the spectrum obtained from the occluded radicals has only one line, whereas a complex hyperfine pattern is obtained in that produced by γ-irradiation.

7.7 RADICALS FORMED BY PYROLYSIS

The fact that large concentrations of free radicals become trapped in organic materials when these are charred appears to have been discovered independently by BENNETT and INGRAM[13]; UEBERSFELD, ETIENNE and COMBRISSON[14], and WINSLOW, BAKER and YAGER[15]. This production and stabilization of free radicals within the carbon ring structure is a very general feature of low temperature pyrolysis, and high radical concentrations have been obtained from all types of charred organic matter[16,17]. A considerable amount of work has now been performed on these radicals by various laboratories; and, although the detailed mechanism of formation and stabilization is not yet unambiguously determined, the experiments have given much information on the different interactions associated with them. These experimental results are first briefly summarized, before the different theories of radical formation are discussed.

The electron resonance absorption obtained from these carbon samples show no resolved hyperfine structure and have *g*–values very close to that of a free-spin. There are therefore only two parameters which vary much between different samples, i.e. the line width and the integrated intensity of absorption. The line widths are found to vary from under 1 gauss to over 100 gauss, and the intensity varies from zero up to values equivalent to 10^{20} free radicals/gramme.

7.7.1 *Variation of Radical Concentration with Carbon Content*

The concentration of free radicals trapped or stabilized in carbonaceous matter is found to increase rapidly as the carbon content rises from 80 per cent to 94 per cent. This is shown in *Figure 59(a)* where the total radical concentration, as obtained from the integrated absorption curve, is plotted against carbon content[18]. X-ray measurements[19,20] indicate that large carbon ring clusters start to

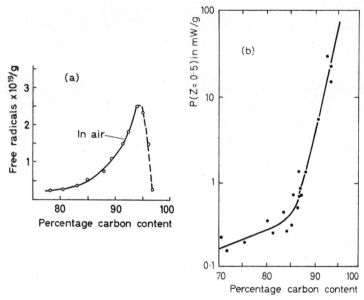

Figure 59. Variation of radical concentration and relaxation time with carbon content. (a) Rise of radical concentration with increasing carbon content[30]; (b) Variation of relaxation time with carbon content, as shown by the saturation parameter. The variable plotted is the power that must be dissipated in the sample for the saturation parameter to be one half (After Smidt[21])

be formed in this same region of carbon content. These results therefore suggest that the stabilization of the radicals may be associated with the high resonance energy available in clusters of four or more benzene rings[18].

More recent measurements by SMIDT[21] and Uebersfeld[42] have shown that chars and coals with low percentage carbon content have electron resonance absorption signals which saturate at relatively low power. These results are summarized in *Figure 59(b)*. In this figure the vertical axis is the power, in milliwatts per gramme, that must be dissipated in the sample for the saturation factor, $Z = [1 + \frac{1}{4} \gamma^2 . H_1^2 . T_1 . T_2]^{-1}$, to be one half. It can be seen that relatively low powers will produce fifty per cent saturation in specimens with 70 to 88 per cent carbon content, whereas no saturation is likely to occur, with the power levels usually available, in specimens with a carbon content above 92 per cent. These results show that care must always be taken to ensure that the measured absorption signals are not being saturated when quantitative readings are required. If the saturation effect had been neglected for the samples and conditions represented by *Figure 59(b)* the measured radical

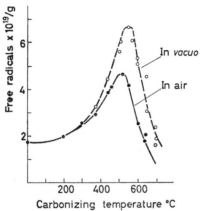

Figure 60. Variation of radical concentration with temperature of carbonization (After Austen, Ingram and Tapley[30])

concentrations would have been too low by a factor of 50 per cent.

These results are also of interest because they show that the spin-lattice relaxation time, T_1, must increase appreciably in the low percentage carbon samples. In the high-percentage carbon samples T_1 has a normal value of about 10^{-7} sec[22], but this increases markedly as the carbon content falls to the 88 per cent point and below. Saturation due to long spin-lattice relaxation time is also often appreciable in the resonance spectra obtained from polymers. It would appear that the random and disorganized structure of these charred or coalified specimens, or of the polymer samples may be a feature which produces small spin-lattice interaction and hence long relaxation times.

7.7.2 *Variation of Radical Concentration with Temperature of Carbonization*

The second parameter to be studied in detail was the effect of carbonization temperature on the free radical concentration [18,23]. The results obtained for a typical sample are shown in *Figure 60,* in which the full curve represents the variation of radical concentration if the samples are observed in air, after cooling, and the dotted curve is for observations *in vacuo*. It will be noted that both curves show a marked rise in the 350° to 550° C region. This is just the region in which most volatile products are driven off and the carbon atoms can be visualised as starting to form condensed ring structures. This again lends support to the postulate that the radicals are stabilized by the formation of resonating ring structures[18,24]. The fall in concentration above 550° C carbonizing temperature, and above 95 per cent carbon content, is probably due to the fact that these rings join to form small graphitic sheets, or cross-link to form a three-dimensional lattice. In either case pairing of their electrons would occur, and the radical concentration would drop.

The difference between the concentrations observed in air and *in vacuo* is very noticeable, and this is due to the interaction of molecular oxygen with the radicals[25, 26]. The experimental observation of this effect is shown in *Figure 61*, and the most striking feature

Figure 61. Effect of oxygen on the electron resonance absorption line. Top left hand trace is that observed in vacuo, and successive traces from left to right were taken at 5 second intervals after initial admission of oxygen (After Ingram and Tapley[25])

is its complete reversibility. The top left-hand trace is the absorption signal observed *in vacuo* from a sample carbonized to about 600° C. The successive traces from left to right across the top and then from left to right across the bottom are photographs of the absorption line taken at 5 second intervals after initial admission of oxygen. This decay of the free-radical signal on contact with molecular oxygen is not so surprising, but the process was found to be completely reversible and it is possible to retrace the whole series of pictures by simply pumping off the oxygen.

7.7.3   *The oxygen effects*

This reversible interaction of the carbon free radicals with molecular oxygen has received considerable attention[25-29]. Two possible mechanisms of interaction between the unpaired electron of the free radical and an oxygen molecule can be postulated. One may be termed ' physical ' and viewed as a form of collision broadening. In this case the oxygen molecule is assumed to approach close to the free radical centre and then move away again, or, alternatively, the free radical electron may move quickly past a stationary oxygen atom. The large disturbance of the electron's energy by the diradical oxygen molecule will reduce the lifetime of its excited state and hence produce a broadening of its energy levels. This may be viewed as an increased spin-lattice interaction effect instead of a form of collision broadening, since both will reduce the characteristic relaxation time. If this ' physical interaction ' is taking place the number of unpaired spins will remain constant, and hence the integrated area of the absorption curve will remain the same, but the peak height will fall as the line broadens out.

The other type of interaction may be termed a ' chemical ' effect, and in this case one of the unpaired spins of the incoming oxygen molecule is assumed to pair with that of the free radical. This ' pairing ' may take place over a relatively long distance, and it is not necessary for a strong covalent bond to be formed. Some type of quasi-chemical bond must be formed, however, so that the unpaired spin of the free radical is no longer effective. If the interaction were with nitric oxide (which also produces the ' oxygen effect ') no unpaired spins would be left and the signal would decay to zero with constant width. Oxygen itself is somewhat more complicated since it possesses two unpaired electrons, and after the formation of this quasi-chemical bond one unpaired spin will still be left localized on an oxygen atom. It will be highly localized, however, and also interacting with any rotational motion of the oxygen molecule, so its absorption spectrum will probably be very anisotropic, and smeared out beyond detection in the amorphous samples that are studied. The net result of oxygen admission would therefore be the same as for the nitric oxide—i.e. a real decay of integrated intensity of the absorption to zero, with constant line width in the process.

Detailed studies on the line shape and intensity of the decaying signals were therefore carried out[27,29] to see if it were possible to differentiate between these two mechanisms of interaction. It was found that, in general, both types of interaction were present since the integrated area is reduced, but the line width also usually increases. AUSTEN and INGRAM[27] were able to find carbon samples showing the two extreme cases, more or less separately, and there seems to be conclusive proof that molecular oxygen interacts with the free radicals both ' physically ' and ' chemically '.

It is interesting in this connection that a reversible oxygen effect has also been observed in some γ–irradiated polymers[9] as discussed in section 7.4. In this case the resonance signal was found to grow on admission of oxygen, and this can only be explained by a chemical interaction. The electron resonance results on these different ' oxygen effects ' have therefore produced conclusive evidence that it is possible for gaseous oxygen to form a quasi-chemical bond with free-radical electrons in which the spins are paired, and it is then possible to break this bond by just reducing the pressure of the oxygen above the sample.

7.7.4 Chemical treatment

The effect of different types of chemical treatment on the radical concentration in these low temperature carbons has also been investigated[24,30]. These experiments were performed to determine whether any particular edge groups were associated with the mechanism of radical formation. Thus the radical concentrations in acetylated coals and their carbonized derivatives have been compared with those in the unacetylated samples to see if the difference in energy required for fission of the C–O–Ac link, as compared with that of the C–O–H link, would cause any change in the carbonization curve. No such change was observed, however, indicating that loss of acetylatable OH groups is not the main mechanism of free radical formation. In a similar way loss of aliphatic hydrogen was shown not to be the main mechanism, and the general conclusion drawn from several similar negative results was that no one edge group in particular is responsible for the radical formation, but that loss of any may be able to liberate electrons into the ring system.

Experiments have also been performed on coals and coal extracts reduced polarographically or by the action of metallic lithium[31] in ethylamine and it is found that there is no decrease in the resonance signal when quinone and similar edge groups are removed. On the other hand a considerable decrease in the free-radical concentration occurs when the reduction of the aromatic rings starts to take place. The general conclusion from all the chemical work is, therefore, that the mechanism of radical formation and stabilization is not associated with any one chemical group in particular, but is probably due to the resonance energy of the condensed carbon rings as a whole.

212

7.7.5 General conclusions

Some quite definite postulates can now be made from the general features of the results summarized above. It has already been seen that the variation of radical concentration with both percentage carbon content and with carbonization temperature suggests that the radical formation and stabilization is associated with the growth of the condensed ring systems. This hypothesis is further supported by the X-ray measurements and results of chemical treatment. It is therefore postulated that the radicals are formed by breakage of the bonds round the edge of the condensed carbon rings, as these grow to form larger systems. In most cases the removal of an edge group from the ring will result in a new C–C bond and thus help the growth of the ring structure. It is postulated, however, that there will be a finite probability for the broken edge-bond to liberate an electron back into the ring system where it will become resonance stabilized. This probability would be expected to rise steeply with increasing ring size. It is also possible that some unpaired electrons may originate from defects in the ring packing, with five or seven-membered rings producing internal trivalent carbon atoms. Either of these two mechanisms envisages the initial rupture of a σ or hyperconjugated bond to liberate the unpaired electron, and then stabilization of this by configurational interaction with the π–bond system. In this way the broken bond must be viewed as having partial σ and partial π orbital character, the latter giving the electron a non-localized and stabilized orbit.

There is evidence for exchange narrowing of the line width at high radical concentrations, and it is thought that the overlap of the electron wave functions probably takes place via the π–orbital systems of several intervening ring clusters. The effect of gaseous oxygen and other paramagnetic entities on the radicals is also evidence for a large amount of delocalization and overlap between the electron orbitals. Measurements have shown that the oxygen only interacts with a small amount of the surface area[32], and since the area occupied by the unpaired electrons is also relatively small, the rapid decay or rise in the free radical signals must be associated with electrons possessing considerable delocalization.

These conclusions can therefore be summarized by saying that the free radicals produced in low temperature carbons appear to be formed by bond breakage at the edge of the condensed carbon rings, they are then stabilized by the large amount of resonance associated with the aromatic rings, and move in highly-delocalized π orbitals over the system.

7.8 ELECTRON RESONANCE IN HIGH TEMPERATURE CARBONS

All the results on the pyrolysed, or charred, carbons summarized above have been for those made below 900° C. It can be seen from *Figure 60* that the radical concentration has a maximum at a carbonizing temperature of about 600° C, and has decreased to a value below detection at 1000° C[15,18]. The carbons formed in the region below 900° C have been termed 'low temperature carbons' and, as seen above, their free radicals are postulated as mobile π electrons stabilized by the aromatic ring clusters in highly delocalized orbitals.

Electron resonance has also been observed in graphite[33,34] and other carbons made above 1400° C. This was first attributed [33,34] to spin resonance from the conduction electrons, but later work indicated that the variation of signal intensity with temperature was not consistent with this view. It was also found[35] that high intensity signals could be obtained from graphitic samples if their surfaces were badly scratched. It would therefore appear that the absorption is actually caused by an electron trapped in a lattice defect or damage centre in the crystal lattice, rather than by an electron or hole in the conduction band. It is more than likely that such structural defects will be formed during the graphitization process, and UBBELOHDE[36] has recently suggested one particular type of defect, known as a 'claw' dislocation, which would have just this ability of holding one unpaired electron.

It would therefore appear that the electron resonance absorption obtained from carbons made above 1400° C arises from defect centres rather than from free radicals or conduction electrons. If preferred they can still be regarded as free radicals, but with a highly localized unpaired electron, confined to the σ orbital of one carbon atom. The high localization of these unpaired spins is confirmed by the fact that they do not interact appreciably with molecular oxygen[37].

One of the most striking features of the study of the radicals trapped in different carbons is the fact that no resonance has been obtained from samples made by prolonged heat treatment in the temperature range 1000°–1400° C. The exact reason for this has still to be found, but a possible explanation may be as follows. As 1000° C is approached all the aromatic ring clusters, formed in the initial low temperature carbonization, may be sufficiently mobile to join and form larger graphitic sheets, or cross link to form a bridging bond in a three-dimensional lattice. In this way the 'low temperature' free radicals will mutually cancel and their concentration is reduced to zero. In the 1000°—1400° C temperature range the system of condensed rings may still retain considerable mobility, as the different planes move into a more parallel orientation. As a result of this mobility no permanent defects in the structure will be produced. As soon as graphitization starts on a large scale, however, the graphitic sheets will become locked in position and 'claw' dislocations and other types of defect will become frozen in the lattice. It would appear that this does not happen to any appreciable extent until 1400° C, and therefore no resonance will be obtained from the lattice defect sites, and localized electrons, until this temperature is reached.

7.8.1 Carbon Blacks

Carbon blacks form rather a special group and can be regarded as a mixture of both 'low and high temperature' carbons. They are usually prepared by very rapid heating in the temperature range 1000°—1700° C, but because the reaction time is so short their electron resonance spectra are more similar to those of the low temperature carbons. Thus, initial measurements[38] on some carbon blacks showed that their resonance also possessed an oxygen effect, and that this could be eliminated on milling with rubber and other long chain polymers. A very systematic study of various carbon blacks has recently been undertaken by COLLINS[37]. He has shown that, when initially formed, their resonance has all the features characteristic of a low temperature carbon. Thus, on reheating slowly, the radical concentration will grow to a maximum at about 700° C, then fall to a very low value and not rise again until the 1400° C temperature is reached. In a similar way carbon blacks that have been re-heated up to 800° C will show an oxygen effect on their resonance signal. It is not until full-length heat treatment above 1400° C has been performed that narrow resonance signals, independent of oxygen absorption, are obtained.

These results are of very great practical significance since they indicate that the free electrons in carbon blacks, as they are normally used commercially, will be mainly in non-localized π orbitals, and not held in lattice defects or traps. Such delocalized radicals may taken an active part in rubber reinforcement and in the other practical applications for which carbon blacks are used.

REFERENCES

[1] FRAENKEL, G. K., HIRSHON, J. M. and WALLING, C. *J. Amer. chem. Soc.* 76 (1954) 3606

[2] SCHNEIDER, E. E., DAY, M. J. and STEIN, G. *Nature, Lond.* 168 (1951) 645

RADICALS FORMED BY POLYMERIZATION AND PYROLYSIS

[3] BAMFORD, C. H., JENKINS, A. D., INGRAM, D. J. E. and SYMONS, M. C. R. *Ibid.* 175 (1955) 894

[4] INGRAM, D. J. E., SYMONS, M. C. R. and TOWNSEND, M. G. *Trans. Faraday Soc.* 54 (1958) 409

[5] ABRAHAM, R. J., MELVILLE, H. W., OVENALL, D. W. and WHIFFEN, D. H. *Ibid.* 54 (1958) 1133

[6] ATHERTON, N. M., MELVILLE, H. W. and WHIFFEN, D. H. *Ibid.* 54 (1958)

[7] SCHNEIDER, E. E. *Disc. Faraday Soc.* 19 (1955) 158

[8] ARD, W. B., SHIELDS, H. and GORDY, W. *J. chem. Phys.* 23 (1955) 1727

[9] ABRAHAM, R. J. and WHIFFEN, D. H. *Trans. Faraday Soc.* 54 (1958)

[10] ATHERTON, N. M., MELVILLE, H. W. and WHIFFEN, D. H. *Ibid.* 54 (1958)

[11] ABRAHAM, R. J. and WHIFFEN, D. H. *Ibid.* 54 (1958)

[12] INGRAM, D. J. E., SYMONS, M. C. R. and TOWNSEND, M. G. (to be published)

[13] — and BENNETT, J. E. *Phil. Mag.* 45 (1954) 545

[14] UEBERSFELD, J., ETIENNE, A. and COMBRISSON, J. *Nature, Lond.* 174 (1954) 614

[15] WINSLOW, F. H., BAKER, W. O. and YAGER, W. A. *J. Amer. chem. Soc.* 77 (1955) 4751

[16] BENNETT, J. E., INGRAM, D. J. E. and TAPLEY, J. G. *J. chem. Phys.* 23 (1955) 215

[17] BENNETT, J. E., INGRAM, D. J. E. and TAPLEY, J. G. *Conf. on Defects in Crystalline Solids.* Bristol 1954. Physical Society, p. 65

[18] INGRAM, D. J. E., TAPLEY, J. G., JACKSON, R., BOND, R. L. and MURNAGHAN, A. R. *Nature, Lond.* 174 (1954) 797

[19] HIRSCH, P. B. *Proc. roy. Soc. A.* 226 (1954) 143

[20] DIAMOND, R. Thesis, 1956, University of Cambridge

[21] SMIDT, J. *Nature, Lond.* 181 (1958) 176

[22] GARIFYANOV, N. S. and KOZYREV, B. M. *J. exp. theor. Phys.* 30 (1956) 1160

[23] INGRAM, D. J. E. *Disc. Faraday Soc.* 19 (1955) 179

[24] AUSTEN, D. E. G. and INGRAM, D. J. E. *Brennstoff Chemie* (1958) s.25

[25] INGRAM, D. J. E. and TAPLEY, J. G. *Chem. & Ind.* (1955) 568

[26] UEBERSFELD, J. and ERB, E. *J. phys. Radium.* 16 (1955) 340

[27] AUSTEN, D. E. G. and INGRAM, D. J. E. *Chem. & Ind.* (1956) 981

[28] UEBERSFELD, J. E. *C. R. Acad. sci., Paris,* 241 (1955) 371

[29] SINGER, L. S. and SPRY, W. T. *Bull. Amer. phys. Soc.* (1956) 214

[30] AUSTEN, D. E. G., INGRAM, D. J. E. and TAPLEY, J. G. *Trans. Faraday Soc.* 54 (1958) 400

[31] — — *Fuel* (1958)

[32] PASTOR, R. C., WEIL, J. A., BROWN, T. H. and TURKEVICH, J. *Phys. Rev.* 102 (1956) 918

[33] CASTLE, J. G. *Ibid.* 92 (1953) 1063; 94 (1954) 1410

[34] HENNIG, G. R., SMALLER, B. and YASAITIS, E. L. *Ibid.* 95 (1954) 1088

[35] MROZOWSKI, S. Report on 3rd International Carbon Conference Buffalo, New York

[36] UBBELOHDE, A. R. *Nature, Lond.* 180 (1957) 380

[37] COLLINS, R. L. Private communication

[38] INGRAM, D. J. E. *Chemisorption.* Keele Conference Report. Butterworths, 1957, p. 260

[39] HARLE, O. L. and THOMAS, J. R. *J. Amer. chem. Soc.* 79 (1957) 2973

[40] BICKEL, A. F. and KOOYMAN, E. C. *J. chem. Soc.* (1956) 2215

[41] BOOZER, C. E., HAMMOND, G. S., HAMILTON, C. E. and SEN, J. N. *J. Amer. chem. Soc.* 77 (1955) 3233, 3238

[42] UEBERSFELD, J. and ERB, E. *C. R. Acad. sci., Paris,* 243 (1956) 2043

8

INORGANIC RADICALS, BIRADICALS AND THE TRIPLET STATE

8.1 Inorganic Radicals

Most free radicals that have any measurable lifetime are associated with organic compounds, because the long carbon chains or aromatic rings give them a certain amount of stability by de-localizing the molecular orbital of the unpaired electron. There are several inorganic radicals of importance, however, and the OH radical which has already been discussed at some length is a good example. Measurements on atomic hydrogen, the simplest radical of all, have also been described in section 6.4.3 and several electron resonance observations have been made on free atoms produced in various ways. In discussing this work it is very difficult to draw a dividing line between 'free radicals' and 'free atoms', and the former term has therefore been taken to include both.

A similar difficulty of nomenclature arises when discussing stable inorganic radicals, such as nitric oxide, chlorine dioxide or the alkali superoxides. These compounds possess an odd number of electrons in their structure, and hence have one unpaired spin and a resulting paramagnetism. Thus, although they themselves are stable compounds under normal conditions, their binding is 'abnormal' compared with the vast majority of compounds which have paired electrons in covalent bonds or in the outer orbitals of ions. In this sense they come under the definition of a free radical given in Chapter 1, and a brief description of their study is included in this chapter for the sake of completeness.

The electron resonance investigations of these inorganic radicals are grouped together by the physical state of the radicals, rather than by their chemical constitution, since the experimental technique and information that can be obtained are determined much more by the former than by the latter. The three groups are as follows:

(i) Gaseous free atoms and radicals.

(ii) Radicals frozen at low temperatures.

(iii) Stable inorganic radicals in the liquid or solid phase.

217

8.2 Gaseous Free Atoms and Radicals

The first electron resonance studies of gases were performed by Beringer and Castle[1]. They employed an H_{011} cylindrical cavity and a bolometer detector, with 30 c/s magnetic field modulation and phase-sensitive detection. This produced a relatively high-sensitivity spectrometer of simple construction with no modulation broadening. They first studied the stable paramagnetic gases, nitric oxide, nitrogen peroxide and oxygen. The simple picture of two electronic energy levels split by the hyperfine interaction, which has served to explain the resonance spectra observed in the solid phase, has to be replaced by a more complex pattern for the gaseous state. This is because there will be strong coupling between the magnetic moment of the unpaired electron and the rotational motion of the molecules, and these interactions will now produce a set of quantized levels, instead of the averaging effect obtained in a liquid sample. The departure of the spectrum and energy level pattern from those of the previous cases will depend on the strength of the coupling to the rotational motion. This is illustrated in the examples summarized below, starting with nitrogen peroxide, which has only a weak coupling, and hence energy levels that can be represented closely by previous equations.

8.2.1 *Nitrogen peroxide,* NO_2

The electron resonance spectra of NO_2 can be explained[2] on the assumption that a Paschen-Back effect has taken place in the fields required for resonance so that the electronic moment, the nuclear moment, and the moment associated with the rotation of the molecule as a whole, are all separately quantized along the direction of the field. The energy levels are then given by the equation

$$E_{M_S, M_I, M_J} = g \cdot \beta \cdot H \cdot M_S + A \cdot M_S \cdot M_I + B \cdot M_S \cdot M_J \quad \ldots (8.1)$$

where M_S is the resolved electron spin quantum number equal to $\pm \frac{1}{2}$; M_I is the resolved nuclear spin quantum number of the nitrogen, equal to $0, \pm 1$; and M_J is the resolved quantum number of the molecular motion. The first two terms are thus similar to those considered in the solid or liquid states, and at 10 mm Hg pressure three overlapping hyperfine lines are obtained. As the pressure is decreased and the collision broadening is reduced these lines are split into a large number of components by the third term, and by the large number of rotational levels which are populated at the temperatures used.

8.2.2 *Nitric oxide*, NO

This case is the opposite extreme to nitrogen peroxide as there is now a strong coupling between the moment of the unpaired electron and the molecular rotation. The $^2\Pi$ ground level of the molecule is split into a lower $^2\Pi_{\frac{1}{2}}$ diamagnetic level and an upper $^2\Pi_{\frac{3}{2}}$ paramagnetic level. This upper level is split into sub-components by interaction with the rotational motion, and transitions between these sub-components were observed by Beringer and Castle[1], and later analysed in more detail by BERINGER, RAWSON and HENRY[3]. The four components of the $J=\frac{3}{2}$ level will produce three absorption lines with the normal $\Delta M_J = \pm 1$ selection rule. These were observed experimentally, and each was further split into three by interaction with the nuclear moment of the N^{14}.

8.2.3 *Oxygen*

Oxygen represents a case half-way between nitric oxide and nitrogen peroxide as the magnetic fields required for resonance are not sufficient to break the coupling between the spin and molecular motion completely and produce a Paschen-Back effect, but appreciable decoupling does occur nevertheless[4]. A detailed theoretical treatment must therefore be made to predict the complete energy level scheme[5-7]. It should be noted that molecular oxygen is in fact a biradical, having two parallel unpaired spins in its $^3\Sigma$ ground state. The orbital moment couples with the spin to give $J=K$, and $K \pm 1$, and the latter are nearly degenerate and separated by about 2 cm^{-1} from $J=K$. It is therefore possible to obtain absorption lines in zero magnetic field at about 5 mm wavelengths. Application of the d.c. magnetic field then splits the J levels, decoupling the K and S vectors as the field rises into the thousand gauss region. A large number of component lines are therefore obtained, but these have been fitted to the theoretical pattern with a very high degree of accuracy[4-7].

The electron resonance measurements on gases are, in fact, a good example of the very precise data that can be obtained on energy level systems and molecular binding. The resonance lines are very sharp when observations are made at low pressures, and the resonance frequencies can therefore be measured to a high order of accuracy. On the experimental side it may be noted that no resolved absorption lines will be observed from oxygen or any other paramagnetic gas if the pressure is much above 20 cm Hg, and the oxygen in a normal cavity at atmospheric pressure will only produce a smeared-out undetectable signal. On the other hand sharp lines due to gaseous oxygen can be obtained in partially evacuated cavities, and these can be mistaken for hyperfine structure components. They can always be identified, however, by the rapid increase of their width with increasing pressure.

8.2.4 *Atomic Hydrogen, Oxygen and Nitrogen*

Beringer and Heald then used the same method that they had employed for the stable paramagnetic gases to investigate free atoms. Atomic hydrogen[8,9] was studied first, and this was produced in a U-shaped discharge tube held directly above the apparatus, and then pumped through a vertical tube leading down the axis of the H_{011} cylindrical resonator. In this way a relatively high concentration of free hydrogen atoms can be kept in the cavity resonator and they were able to make very precise measurements on the exact g-values and hyperfine splittings. The same method was then applied to atomic oxygen[10]. Six electronic transitions are observed in this case and these arise from both the $^3P_2$ ground state and the $^3P_1$ metastable state of the oxygen atom. Three extra lines were also observed interleaving the four $^3P_2$ transitions, and these were later identified as double-quantum transitions by HUGHES and GEIGER[11]. Atomic nitrogen has also been studied in the same way[12], this exists in a ^4S ground state, and a hyperfine splitting from the N^{14} nucleus is obtained.

8.2.5 *Atomic Phosphorus*

The free hydrogen, oxygen and nitrogen atoms studied by BERINGER *et al.*[9,10,12] were relatively easy to produce by passing a high voltage discharge through the wet molecular gas at 0·1 mm Hg pressure. Recombination on the glass walls could be prevented by coating them with a suitable anticatalyst, and relatively high concentrations of free atoms could thus be obtained in the resonator. Unfortunately the mechanism of the wall poisoning by the anti-catalysts is not well understood, and suitable compounds for other free atoms have not yet been found. DEHMELT[13] therefore proposed that a more general technique of dissociation could be adopted in which the active component at a pressure between 0·01 and 0·1 mm Hg is added to an inert gas carrier at a comparatively high pressure of 10–100 mm Hg. At these pressures it is possible to maintain high temperature arcs with gas temperatures up to 6,000° K. It is therefore possible to dissociate any gas molecule by purely thermal means, and at the same time the high pressure carrier gas greatly reduces the diffusion speed of the free atoms and thus nearly eliminates the wall recombination effects. Dehmelt[13] tried out these ideas experimentally and used a mixture of helium and argon as the inert carrier gas. He first showed that atomic nitrogen could be produced in this way and the same spectrum was obtained as by HEALD and BERINGER[12]. He then employed the method to study atomic phosphorus vapour. In this case the inert helium-argon mixture was first passed into a flask containing white phosphorus, and the mixture was then fed through the high voltage discharge tube. The gas stream was pumped continuously from here through a tube down the axis of an H_{012} rectangular cavity, and in this way a concentration of about 10^{14} free phosphorus atoms could be maintained in the cavity. The phosphorus is in a $^4S_{\frac{3}{2}}$ ground state and its resonance spectrum is split into a doublet of 20·0 gauss separation, due to the nuclear spin of $I=\frac{1}{2}$ of P^{31}.

8.2.6 *Atomic Iodine*

The use of electrical discharges is not the only method whereby free atoms can be produced in the cavity resonator. A more general method is that of photodissociation, and BOWERS, KAMPER and KNIGHT[14] constructed a high sensitivity X-band spectrometer to investigate free atoms produced in this way. Considerable care was taken in the stabilization of the magnetic field and microwave frequencies, and by using high frequency field modulation with phase-sensitive detection a spectrometer with high sensitivity and with precise frequency and field measurement was obtained. The u.v. beam was focussed into the H_{011} cylindrical cavity through a central hole of 1 cm diameter in the top end. The first systematic measurements on photodissociated atoms were those on iodine vapour[15], and six of the eighteen $\Delta M_I=0$, $\Delta M_J=1$ transitions of the $^2P_{\frac{3}{2}}$ ground state of atomic I^{127} were measured with an accuracy of one or two parts per million.

It was also possible to measure the dependence of the free atom concentration on the different irradiation conditions. It was found that if the partial pressure of the molecular iodine were held constant at about 0·1 mm and the total pressure increased by the addition of air, the concentration of free atoms remained constant. It was also insensitive to changes in the partial pressure of the molecular iodine in the range between 0·02 mm and 0·2 mm Hg. It therefore follows[15] that the rate of recombination, and the rate of dissociation by light in the banded region of the spectrum, are approximately independent of pressure. Moreover, it would appear that the only recombination mechanism which can account for all the observations is a three-body collision in the gas phase with an iodine molecule acting as the third body.

8.2.7 *Summary*

It can be seen from this brief description of electron resonance studies on paramagnetic gases and free atoms that quite a variety of different species can be investigated by this means. In the experiments so far performed most of the effort has been to obtain a detailed picture of the energy level scheme within the free atoms, rather than to study their active chemical properties. The work on the iodine atoms has shown, however, that considerable information of the latter kind can also be obtained, and it is for this reason that the practical experimental details, instead of the precise quantitative results, have been summarized in the above sections. Experiments on more complex gaseous free-radical species, employing the same kind of technique, will undoubtedly follow in the near future.

There is one practical point of some importance that should be mentioned. This is the fact that it is very necessary to prevent free electrons from entering the cavity when the free atoms or free radicals are being studied. Any free electrons will interact with the microwave electric field and undergo a cyclotron resonance as they absorb energy into their orbital motion[16]. This absorption couples via an electric dipole and has a very much higher transition probability than the magnetic dipole coupling of the normal electron resonance transition. As a result a large absorption of considerable width will be produced by a small number of free electrons in the cavity, and this may obscure the spectra from any free atoms or radicals that are also present.

8.3 RADICALS FROZEN AT LOW TEMPERATURES

It is possible to condense the products of a gaseous discharge at very low temperatures and trap the free atoms that are formed, in a matrix of frozen molecules. This technique is somewhat similar to the low temperature trapping techniques discussed in Chapter 6, but in this case the free atoms or radicals are actually present as the material is deep frozen. JEN, FONER, COCHRAN and BOWERS[17] studied hydrogen and deuterium atoms trapped in this way, and a brief description of their apparatus will illustrate the techniques involved. The free atoms were produced in an electrodeless discharge at 0·1 mm Hg pressure excited by a 100 watt 4 Mc/s radiofrequency oscillator. The products of the discharge were pumped past a short side tube which was connected to the low temperature cell. This tube was terminated by a narrow slit which formed a molecular beam of the discharge tube products. A sapphire rod, cooled to 4° K, was held at a distance of 3 cm from the slit so that the molecular beam fell on this and its constituents were immediately deep frozen. This deposit was allowed to accumulate for some minutes and the sapphire rod was then inserted centrally down the axis of an H_{011} cylindrical cavity. Spectra of atomic hydrogen or deuterium were obtained, and these could be identified as due to the free atoms by their hyperfine splittings. The water vapour discharge gave two lines, each of 0·7 gauss width, and of 508·7 gauss separation, and this splitting is nearly the same as that of free gaseous hydrogen atoms. Similarly, a heavy water discharge produced a deposit which had a triplet resonance spectrum, with a splitting of 76·7 and 78·7 gauss separation between the lines. The precise resonance frequencies were determined and the Breit-Rabi hyperfine interaction equation was applied to them. In this way it was shown that the trapped atoms were very nearly, but not exactly, free. It can be seen that these experiments are very similar to those of LIVINGSTON, ZELDES and TAYLOR[18] on γ-irradiated frozen solutions, summarized in section 6.4.3. Another feature in common was the fact that the weak satellite lines due to coupled proton transitions[19] were also observed on the spectra of the frozen discharge products. The free hydrogen and deuterium atoms trapped in this way were found to be quite stable as long as the deposit was kept at 4° K. Precise details of the apparatus required to carry out these very low temperature trapping techniques can be found in a review paper by BROIDA[20].

Additional resonance lines not due to the atomic hydrogen were also observed in the deposits at 4° K and these were probably due to OH· or HO_2^- radicals. LIVINGSTON, GHORMLEY and ZELDES[21] studied the products of electrical discharges when condensed at 77° K and obtained a wide asymmetrical line. The same line shape was observed from the frozen deposits if heavy water or hydrogen peroxide were used in the discharge instead of ordinary water vapour, and also if hydrogen or ammonia were first passed through the discharge and then blended with oxygen. All the samples prepared in this way had a pale yellow colour and both the colour and electron resonance signals disappeared when the samples were warmed to −135° C. Measurements at different frequencies confirmed the fact that the asymmetry of the line shape was due to an anisotropic g value, and it would appear that the trapped species are therefore OH· or HO_2^- radicals. The absorption line is in fact very similar to that observed from NaO_2 in the solid state[23,24], and this suggests that HO_2^- radicals are likely to be the actual entity trapped. It is noticeable that there was no trace of atomic hydrogen in these deposits formed at 77° K, and it is likely that the

223

colder helium temperatures are necessary to prevent diffusion of the free atoms.

The electron resonance spectrum from the products of a nitrogen glow discharge, condensed at 4° K, has also been studied by COLE, HARDING, PELLAM and YOST[22]. In their experiments the discharge through the nitrogen gas was adjusted until the characteristic yellow afterglow was obtained, and the glowing gas was simply condensed on the walls of the cavity resonator, which was cooled in liquid helium. A well-resolved triplet structure was obtained on the electron resonance signal, with a splitting between the lines slightly greater than that obtained by Heald and Beringer[12] for gaseous free nitrogen atoms. This would appear to be conclusive evidence that free nitrogen atoms have been trapped, but more recent results[55] show that the nature of the molecular matrix affects the splitting.

The general conclusion from these preliminary measurements on deep frozen deposits from gas discharges is that temperatures in the liquid helium region are necessary if free atoms are to remain trapped in the frozen molecular matrix, but that free radicals containing more than one atom, can be trapped at liquid nitrogen or oxygen temperatures. Since it should be possible to form and trap a large variety of different radicals in this way, this particular technique will probably be used to a considerable extent in future free-radical studies.

8.4 STABLE INORGANIC RADICALS IN THE LIQUID OR SOLID PHASE
There are a few inorganic compounds which possess an odd number of electrons associated with their binding orbitals, and which are in a solid or liquid phase under normal conditions of temperature and pressure. These are distinct from the transition group compounds, in which the unpaired electrons are located in inner unfilled shells, and are a property of the paramagnetic atom and not of the molecular binding. Examples of the former compounds are the alkali metal superoxides NaO_2 and KO_2, their ozonates NaO_3 and KO_3, and chlorine dioxide, ClO_2, which liquefies at a temperature not far below 0° C. It will be seen that each of these molecules has an odd number of total electrons, and since the constituent atoms have no unfilled lower electron orbits, there must be an unpaired electron in the bonding orbitals. As such the compounds come under the general definition of ' free radicals ' but they are not active radicals in the chemical sense, since they have no great affinity for combination with other groups. Their chemical binding is sufficiently interesting to merit some consideration, however, and their electron resonance spectra are summarized briefly below.

8.4.1 *The alkali superoxides*

Sodium and potassium superoxides form a crystal lattice composed of Na^+ or K^+ and O_2^- ions. The unpaired electron thus resides on the O_2^- ions and can be viewed as moving in a molecular orbital around the two oxygen atoms. The electron resonance spectra observed[23] in these compounds consists of one asymmetrical line, and the separation between its high-field maximum and low-field shoulder is directly proportional to the resonance frequency. This shows that the shape of the absorption line is due to a g-value variation and the measured values are $g_\| = 2\cdot175$ and $g_\perp = 2\cdot002$, the $g_\perp$ being associated with the maximum, and the $g_\|$ with the lower field shoulder.

The electronic structure of the O_2^- ion is very similar to that of nitric oxide but has the $^2\Pi_{\frac{3}{2}}$ and $^2\Pi_{\frac{1}{2}}$ levels inverted so that the former is the ground state. In the crystalline phase the symmetry about the molecular axis is removed and one of the $\pi_g 2p$ orbitals will have higher energy than the other. The measured g-values can then be employed to determine the ratio of the splitting between these two levels to the spin-orbit coupling constant[23], and this ratio will, in turn, determine the other magnetic properties of the complex.

The electron resonance absorption lines obtained from the ozonates[24] are much more symmetrical, and the shoulder now occurs on the high field side. Measurements at different wavelengths again confirm that the asymmetry is due to a true g-value variation and the values obtained are $g_\| = 2\cdot003$ and $g_\perp = 2\cdot015$.

Electron resonance absorption has also been obtained from sodium dithionite[54], and it is thought that this may be due to SO_2^- radicals. One interesting feature of these results is that the largest free radical concentration is obtained when there is a liquid-solid interface present in the specimen. This suggests that the radicals may be produced in the liquid phase but are stabilised by the presence of the solid surface. Unlike the other sulphur-containing radicals there is no evidence for an anisotropic g-value or one displaced much from the free-spin value.

8.4.2 *Chlorine dioxide*

Electron resonance spectra can be obtained from liquefied chlorine dioxide and from solutions of the gas in water, acetone, benzene and other organic solvents[25]. A single broad absorption line of 150 gauss half width is obtained from the pure chlorine dioxide in either the liquid or solid phase, or from concentrated solutions of the gas. If dilute solutions are studied, however, a well resolved four-component hyperfine pattern is observed, due to the interaction with the chlorine nuclei (both of spin $I = \frac{3}{2}$). The separation between the individual components is 17 gauss and each line has a width of 8 gauss. These narrow well resolved lines are produced by the averaging effect of the liquid tumbling motion, which reduces the anisotropic hyperfine contributions to zero, as in the case of the solutions of organic free radicals. Experimental confirmation of this fact is obtained in a striking manner by simply freezing the solutions when a much wider and less-resolved pattern is produced[24]. This can be analysed as an integrated overlap of lines with a hyperfine splitting parameter varying from 21 gauss to 52 gauss with a corresponding g-value variation from 2·02 to 2·01. These experiments thus afford a very direct example of the difference between the anisotropic hyperfine interaction, which is averaged to zero in the liquid state, and the isotropic term which is not affected by the averaging process.

These results have also been analysed theoretically[24] and they show that the orbit of the unpaired electron is very largely centred on the Cl nucleus, and has approximately 70 per cent $3d$ character and 30 per cent $3p$ character.

8.4.3 *Metal ammonia solutions*

Another inorganic system in which an unpaired electron is found in an outer orbital of an atom is that of the alkali metal ammonia solutions. These again are somewhat of a specialized class of compounds and would not normally be classified as ' free radicals '. The alkali metals ionize in such solutions to give positive metal ions with full closed shells and no unpaired electrons, together with negative ions which can be visualized as electrons moving in a cavity formed by the surrounding ammonia molecules. HUTCHISON and PASTOR[26] were the first to observe electron resonance signals from such solutions and very narrow absorption lines with g-values very close to that of a free spin were obtained. It was also shown[27,28] that the same spectra could be obtained with other solvents, such as methylamine and ethylenediamine. A series of systematic experiments[29] on sodium and potassium in different solvents and at different concentrations showed that there was a minimum line width of 0·02 gauss. The g-value was independent of concentration and equal to $2·0012 \pm 0·0002$. The theoretical implications of the model, in which the electrons are visualized as located in cavities in the ammonia solution, have been investigated by KAPLAN and KITTEL[30]. The electron is considered as moving in a molecular orbital spread over several protons of the neighbouring ammonia molecules. At sufficient dilution the electron spin-spin broadening will be negligible but a width due to unresolved hyperfine structure from the protons is to be expected. On substituting the approximate conditions expected for such cavities it is found that the envelope of such unresolved hyperfine structure should have a width of about ten gauss. This broadening will be considerably reduced by the rapid motion of the ammonia molecules, however, because the unpaired electron will only interact with a given proton for a fraction of the expected time. As a result of this the predicted line width is reduced to 0·05 gauss which is in very good agreement with that observed experimentally.

8.4.4 *Other inorganic systems possessing free electrons*

There are several other inorganic systems which contain free electrons in their structure, such as metals with their conduction electrons, and semi-conductors with their current carriers. These cannot be classified as free radicals, however, under any form of definition and so they are not considered in this book. It may be noted nevertheless, that electron spin resonance has been used extensively to study both the conduction electrons in metals[31-36] and the impurity centres in semi-conductors[37-38]; and cyclotron resonance has also been employed to study the properties of the current carriers in semi-conductors[39-42]. Further details on this work can be found in the references indicated.

8.5 BIRADICALS

It is possible for a molecule to have two unpaired electrons associated with its chemical binding, and in this case it is termed a bi- or diradical. One case has already been considered above, namely, that of the oxygen molecule, and it was seen there that the two unpaired spins coupled and aligned in this instance to give a ground state of $^3\Sigma$. The coupling between the two unpaired electrons can vary very appreciably, from zero to that of a strong interaction, and the nature and properties of the biradicals will vary accordingly.

Two extreme cases may be taken to illustrate this point. First that of a long polymer chain which has an active radical at each end, each with an unpaired electron (as might be formed by X- or γ-irradiation). These two electrons will be very far apart and, as seen in Chapter 7, their molecular orbits will not embrace more than three or four groups at either end. There will therefore be very little, or no, interaction between the two, and each will behave separately as a single free radical. Their electron resonance spectrum will thus be identical with that obtained from the mono-radical, but twice as intense per polymer chain length.

As the other extreme case, a photo-sensitive molecule such as fluorescein may be taken. In this case the molecule will normally have no unpaired spins. On irradiation, however, a molecular bond is broken and one of the electron spins reverses sign so that the two electrons are now aligned with their spins parallel. These two electrons will be interacting very strongly, and the two spins will couple to form a resultant vector of $S=1$. This total spin will thus be able to take up three, i.e. $(2S+1)$, different orientations in an applied magnetic field and so produce three different energy levels. This state of the molecule is therefore often referred to as ' the triplet state'. It may also be classified as a ' biradical' since the molecule now possesses two unpaired electrons, but in such cases of strong coupling the term ' excitation to a triplet state' is usually employed.

There will, of course, be a whole range of intermediate couplings between these two extreme cases. An example of such is Chichiba-bin's hydrocarbon, p,p-biphenylene-bis-diphenylmethyl, which has the following structural formula, when in its normal diamagnetic state

It is possible, however, for the molecule to be excited so that the electrons in one of the double bonds unpair, and the conjugation of the central portion changes to give a structure of the form

This molecule now possesses two unpaired electrons which may be visualized as centred on carbon atoms at a relatively large distance apart. The molecule can thus justifiably be termed a biradical. On the other hand, the excitation from the state with no unpaired electrons (singlet state) to the state with the two unpaired electrons (triplet state) is produced by a simple rearrangement of the conjugated bond system, and the change can therefore also be justifiably termed a ' triplet state excitation '.

It will be seen in the electron resonance results summarized below, however, that experimentally the compounds studied fall into two clear cut groups. On the one hand detailed measurements have been made on molecules containing two unpaired electrons separated by several ring systems, and very little interaction is then found between them. On the other hand attempts have been made to pick up the resonance from triplet states produced by irradiation of photosensitive molecules, in which strong coupling between the electrons is known to exist. The term ' biradicals ' will be used in the rest of the book to describe the former type of molecule, and the term ' triplet state ' to describe the latter.

It may be noted that no conclusive experiments have yet been performed on the inte mediate compounds, such as Chichibabin's hydrocarbon. The magnetic susceptibility measurements[43] show that this particular molecule is diamagnetic and hence the amount of triplet state excitation must be small, even after correcting for possible errors in the values assumed for the molecular diamagnetism[44]. Electron resonance signals have been obtained[45] from its solutions which might be interpreted as evidence of 4 per cent biradical formation, but there is a distinct possibility[46] that the observed signal may have been due to an impurity. Another case similar to Chichibabin's hydrocarbon is that of bianthrone. The electron resonance measurements[47] on this molecule have been summarized in section 5.3.2 and it was seen there that a biradical positive ion was one possible form for the paramagnetic compound. There are several other monoradical forms that could be produced as well, however, and there is no conclusive evidence for a triplet state in bianthrone. In the results summarized in the section below it will be seen that all the biradicals which give definite electron resonance spectra have only weakly coupled electrons, as shown by very small shifts of the g-value and the absence of electronic splittings.

8.6 EXPERIMENTAL RESULTS ON BIRADICALS

The first systematic series of experiments on different organic biradicals were performed by JARRETT, SLOAN and VAUGHAN[48].

They studied three general series of compounds which were:—

(*i*) The 4,4'-polymethylenebistriphenylmethyl radicals of the general structural formula

with *n* varying from 1 to 4.

(*ii*) The p-substituted polyphenyls, of the general structural formula

with *n* equal to one or two, and with different groups at R_1 and R_2.

(*iii*) Oxybistriphenylmethyl, with the structural formula

It will be seen that a large number of different groups had been substituted between the carbons localizing the unpaired electrons, when these measurements were completed. One striking result was that all the measured *g*-values were very close to the free-spin value. This shows that any coupling between the unpaired spins must be small, since the spin-orbit coupling would split the three levels of a triplet state[49], formed from aligned spins, to give effective *g*-values displaced from 2·0.

The spectra obtained from solutions of the first group of compounds in benzene are shown in *Figure 62*. A well-resolved hyperfine splitting is obtained when there is only one CH_2 group in the central chain. This arises from the protons in the two phenyl and one

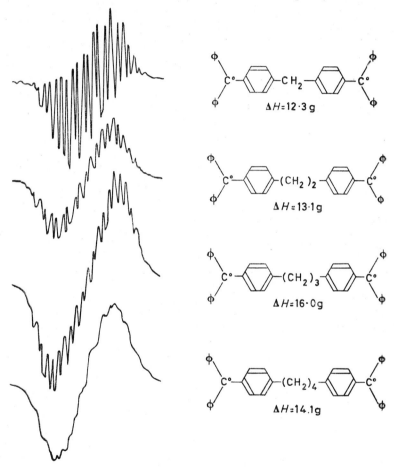

Figure 62. Hyperfine structure in biradical spectra of polymethylenebistriphenylmethyl radicals (After Jarrett, Sloan and Vaughan[48])

phenylene group which are embraced in the molecular orbit of the unpaired electron, in a way very similar to that in triphenylmethyl. The phenylene groups are separated by one or more methylene groups and hence little interaction between the unpaired

231

electrons is to be expected. This is confirmed by the fact that the g-value was $2 \cdot 0025 \pm 0 \cdot 0004$ and thus very close to the $2 \cdot 0023$ of a free spin, for all four compounds. It is seen, however, that the resolution of the hyperfine patterns becomes progressively worse as the chain becomes longer, and this is probably due to an increasing restriction on the tumbling motion and hence a less effective averaging of the anisotropic hyperfine interaction.

The hyperfine patterns obtained from the second group of compounds were not so well resolved, which was probably due to the fact that larger substituent groups were used at R_1 and R_2 and more overlap of the resulting proton interactions resulted. The g-value of these compounds was again found to be equal to $2 \cdot 0025$ showing that there was very little interaction between the unpaired electrons.

The hyperfine pattern of the 4, 4' oxybistriphenylmethyl was well resolved and very similar to that of the methylene derivative shown at the top of *Figure 62*. It would appear that the molecular orbit of the electron again embraces the two phenyl and one phenylene group, as before. The g-value has risen slightly in this case, however, to $2 \cdot 0031 \pm 0 \cdot 0004$ indicating that there is a slightly larger spin-orbit coupling, and interaction between the two electrons. This might be expected since there is now a π orbital on the oxygen which can couple across between the two phenylene groups. The g-value shift is still very small, however, and the general conclusion of these measurements is that the unpaired electrons in such biradicals as these behave more like two uncoupled mono-radicals, than two coupled electrons giving a total S of one and a triplet energy state.

A very striking confirmation of the weak coupling and slow exchange between the two electrons has recently been obtained by REITZ and WEISSMAN[50]. They studied the spectrum of the 4, 4' oxybistriphenylmethyl radical with C^{13} substituted into the methyl position in different abundances. If one of the methyl carbons is C^{13} and the other is C^{12}, the hyperfine pattern observed from the C^{13} will vary according to the frequency at which the electron spins exchange between the two atoms (which is a measure of the coupling between them). If the spin exchange frequency is slow, compared with the hyperfine frequency, the C^{13}–C^{12} group will give a triplet splitting with the central line, which comes from the C^{12}, twice as intense as the two outer components, which originate from an equal amount of C^{13}. If the spin exchange frequency is higher than the hyperfine frequency, however, the two electrons can be considered as coupled to form a resultant $S=1$ which interacts simultaneously with both nuclei. Hence only the two outer lines will be observed and the central line will be absent. Similar reasoning can be applied to the C^{13}–C^{13} case when two lines are obtained for slow spin-exchange but a triplet for fast spin-exchange. The hyperfine pattern obtained by Reitz and Weissman[50] showed conclusively that the spin exchange was slow compared with the hyperfine splitting, and must be less than 10^8 c/s. These results are of considerable interest as introducing a new quantitative method for the determination of electron interaction in a biradical or triplet state.

It should be noted that a biradical complex can also be formed by two unpaired electrons on different molecules. If these approach sufficiently close to one another their spins may couple and a hyperfine pattern equivalent to equal interaction with the total number of protons in both groups may be obtained. This possibility was first suggested by SCHNEIDER[51] for the case of polymer chains severed by γ-rays. Although the explanation is probably not correct in this particular case, a similar interaction may well account for the five-line spectra observed when two R–CH_2 radicals approach[52], as discussed in more detail in section 6.2.4.

8.7   THE TRIPLET STATE

It has been seen in the above section that all biradicals, for which electron resonance spectra have been observed, have unpaired electrons which are only weakly coupled, and act more as separate mono-radicals than a triplet system. There are certain complexes, however, in which it is known that strong coupling does exist between the electrons, and a true ' triplet excitation ' is produced. Photo-sensitive fluorescent compounds are one such group, and the paramagnetism of such molecules excited to the triplet state can be detected by magnetic susceptibility measurements[53]. It was therefore expected that electron resonance techniques would be able to detect these excited triplet states quite easily, since the sensitivity available is some orders of magnitude greater than that of the susceptibility measurements. A large number of different laboratories have set up equipment to perform such experiments and it is probably fair to say that the complete failure to observe triplet state excitation has been one of the biggest disappointments in electron resonance studies. Various workers have tried high intensity illumination of such compounds as fluorescein, anthracene, acridine orange, dibenzothiophene, and triphenylene, under conditions in which it is known that at least 10^{14}–10^{15} unpaired electrons will be present. In no case has there been conclusive evidence for electron resonance absorption by a triplet state. This therefore indicates that some factor must always be entering to broaden the signal beyond the limits of detection.

A consideration of the likely behaviour of the triplet energy levels in any non-crystalline solid material will show, in fact, that

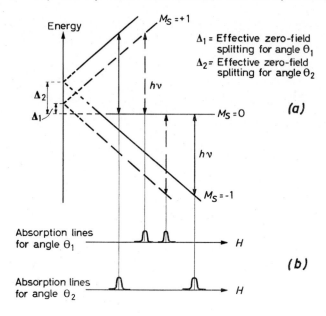

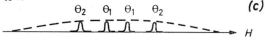

Figure 63. *Energy level splitting of a triplet state.* (a) *Level diver-gence for two different angles between applied magnetic field and mole-cular axis, showing variation of effective zero field splitting;* (b) *Expected absorption lines for a given frequency;* (c) *Smeared out spectrum obtained from polycrystalline or amorphous sample. It should be noted that the actual energy level divergence is not linear at low field values*

this is to be expected. It has already been seen that there will be a significant spin-orbit interaction if the electrons couple to form a resultant spin of $S=1$. This interaction will split the three levels of the triplet state even in zero magnetic field, and in any molecule without spherical symmetry the effect of this splitting on the spectrum will vary with the angle between the applied magnetic field and the molecular axis. This variation in splitting between the levels, on application of a magnetic field, arises from the two competing axes of quantization, that of the magnetic field, and that of the molecular axis of symmetry. Its effect on the resultant change of energy levels with field is summarized in *Figure 63* for two different angles. It is seen from this that not only are two separate absorption lines now produced as a result of the zero-field splitting, but also that the separation between these varies with angle. Most of the compounds which have been irradiated to produce triplet excitation have been held in silica gel, boric acid or low temperature glasses, and their molecules thus have a random orientation with respect to any applied external field. It therefore follows that their spectra will be a series of overlapping doublets of varying separation, and hence the envelope may be broadened well beyond detection.

It would therefore appear that triplet state excitation will only be observed by electron resonance if single crystals are studied. In these the splittings should be the same for each molecule and identical overlapping doublets should be produced. This is not the complete answer to the problem, however, since single crystal irradiation has already been attempted, with this in mind, but so far no spectra have been observed. It is likely that another major contribution to the broadening will be the spin-lattice relaxation time. If there is strong interaction with the lattice via the triplet excitation very low temperatures may have to be employed before sufficiently narrow lines are obtained. On the other hand, if a metastable state produces relative isolation of the spins from the lattice, saturation effects may occur, and low microwave power levels may have to be employed. It would appear in fact that considerable care will have to be taken in choosing exactly the right conditions for the observation of triplet state excitation, and this remains as one of the major challenges to electron resonance research workers.

REFERENCES

[1] BERINGER, R. and CASTLE, J. G. *Phys. Rev.* 78 (1950) 581
[2] CASTLE, J. G. and BERINGER, R. *Ibid.* 80 (1950) 114
[3] BERINGER, R., RAWSON, E. B. and HENRY, A. F. *Ibid.* 94 (1954) 343
[4] —and CASTLE, J. G. *Ibid.* 81 (1951) 82
[5] HENRY, A. F. *Ibid.* 80 (1950) 396
[6] TINKHAM, M. and STRANDBERG, M. W. P. *Ibid.* 97 (1955) 937, 951
[7] — — *Ibid.* 99 (1955) 537
[8] BERINGER, R. and RAWSON, E. B. *Ibid.* 87 (1952) 228
[9] —and HEALD, M. A. *Ibid.* 95 (1954) 1474
[10] RAWSON, E. B. and BERINGER, R. *Ibid.* 88 (1952) 677
[11] HUGHES, V. W. and GEIGER, J. S. *Ibid.* 99 (1955) 1842

[12] HEALD, M. A. and BERINGER, R. *Ibid.* 96 (1954) 645
[13] DEHMELT, H. G. *Ibid.* 99 (1955) 527
[14] BOWERS, K. D., KAMPER, R. A. and KNIGHT, R. B. D. *J. sci. Instrum.* 34 (1957) 49
[15] —— and LUSTIG, C. D. *Proc. phys. Soc. B* 70 (1957) 1177
[16] INGRAM, D. J. E. and TAPLEY, J. G. *Phys. Rev.* 97 (1955) 238
[17] JEN, C. K., FONER, S. N., COCHRAN, E. L. and BOWERS, V. A. *Ibid.* 104 (1956) 846
[18] LIVINGSTON, R., ZELDES, H. and TAYLOR, E. H. *Disc. Faraday Soc.* 19 (1955) 166
[19] ZELDES, H. and LIVINGSTON, R. *Phys. Rev.* 96 (1954) 1702
[20] BROIDA, H. P. *Ann. Acad. Sci., N.Y.* 67 (1957) 530
[21] LIVINGSTON, R., GHORMLEY, J. and ZELDES, H. *J. chem. Phys.* 24 (1956) 483
[22] COLE, T., HARDING, J. T., PELLAM, J. R. and YOST, D. M. *Ibid.* 27 (1957) 593
[23] BENNETT, J. E., INGRAM, D. J. E., SYMONS, M. C. R., GEORGE, P. and GRIFFITH, J. S. *Phil. Mag.* 46 (1956) 443
[24] —— and SCHONLAND, D. *Proc. phys. Soc. A* 69 (1956) 556
[25] —— *Phil. Mag.* 1 (1956) 109
[26] HUTCHISON, C. A. and PASTOR, R. C. *Phys. Rev.* 81 (1951) 282
[27] GARSTENS, M. A. and RYAN, A. H. *Ibid.* 81 (1951) 888
[28] LEVINTHAL, E. C., ROGERS, E. H. and OGG, R. A. *Ibid.* 83 (1951) 182
[29] HUTCHISON, C. A. and PASTOR, R. C. *J. chem. Phys.* 21 (1953) 1959
[30] KAPLAN, J. and KITTEL, C. *Ibid.* 21 (1953) 1429
[31] GRISWOLD, T. W., KIP, A. F. and KITTEL, C. *Phys. Rev.* 88 (1952) 951
[32] SOLT, I. H. and STRANDBERG, M. W. P. *Ibid.* 95 (1954) 607
[33] KIP, A. F., GRISWOLD, T. W. and PORTIS, A. M. *Ibid.* 92 (1953) 544
[34] CARVER, T. R. and SLICHTER, C. P. *Ibid.* 92 (1953) 212
[35] FEHER, G. and KIP, A. F. *Ibid.* 95 (1954) 1343; 98 (1955) 337
[36] GUTOWSKY, H. S. and FRANK, P. J. *Ibid.* 94 (1954) 1067
[37] FLETCHER, R. C., YAGER, W. A., PEARSON, G. L., HOLDEN, A. N., READ, W. T. and MERRITT, F. R. *Ibid.* 94 (1954) 1392
[38] —— and MERRITT, F. R. *Ibid.* 95 (1954) 844
[39] DRESSELHAUS, G., KIP, A. F. and KITTEL, C. *Ibid.* 92 (1953) 827; 98 (1955) 368
[40] DEXTER, R. N., ZEIGER, H. J. and LAX, B. *Ibid.* 95 (1954) 557
[41] DINGLE, R. B. *Proc. roy. Soc. A* 212 (1952) 38
[42] SHOCKLEY, W. *Phys. Rev.* 90 (1953) 491
[43] MÜLLER, E. and MÜLLER-RODLOFF, I. *Ann.* 517 (1935) 134
[44] SELWOOD, P. W. and DOBRES, R. M. *J. Amer. chem. Soc.* 72 (1950) 3860
[45] HUTCHISON, C. A., KOWALSKY, A., PASTOR, R. C. and WHELAND, G. W. *J. chem. Phys.* 20 (1952) 1485
[46] HUTCHISON, C. A. Private communication
[47] HIRSHON, J. M., GARDNER, D. M. and FRAENKEL, G. K. *J. Amer. chem. Soc.* 75 (1953) 4115
[48] JARRETT, H. S., SLOAN, G. J. and VAUGHAN, W. R. *J. chem. Phys.* 25 (1956) 697
[49] COULSON, C. A. *Valence.* Oxford University Press, 1952, p. 137
[50] REITZ, D. C. and WEISSMAN, S. I. *J. chem. Phys.* 27 (1957) 968
[51] SCHNEIDER, E. E. *Disc. Faraday Soc.* 19 (1955) 158
[52] INGRAM, D. J. E. and FUJIMOTO, M. *Trans. Faraday Soc.* 54 (1958)
[53] EVANS, D. F. *Nature, Lond.* 176 (1955) 777
[54] HODGSON, W. G., NEAVES, A. and PARKER, C. A. *Nature, Lond.* 178 (1956) 489
[55] FONER, S. N., JEN, C. K., COCHRAN, E. L. and BOWERS, V. A. *J. chem. Phys.* 28 (1958) 351

BIOLOGICAL AND MEDICAL APPLICATIONS

9.1 THE DIFFERENT APPLICATIONS OF ELECTRON RESONANCE

BEFORE considering the application of electron resonance to biological and medical problems in particular, a brief summary is first given of its different uses in the study of physical and chemical systems in general. The results which have been discussed in the preceding chapters may be taken as a basis for such a summary, and the different applications can be listed as follows.

(i) *The quantitative determination of free radical concentration*

It was seen in section 3.6 that the minimum number of free radicals that can be detected under optimum conditions in an X-band spectrometer is given by $10^{11}.\Delta H.\tau^{-\frac{1}{2}}$ where ΔH is the width of their absorption line in gauss, and τ is the time taken to sweep through the resonance. Most free radicals occurring in biological processes have a line width of about 10 gauss, and a typical value of τ used experimentally is 10 seconds. This gives a minimum detectable free radical concentration of 3.10^{11} per gramme (10^{-9} molar) under optimum conditions. It should be noted, however, that the presence of water, which has a high non-resonance absorption, may reduce the sensitivity considerably. If steady state radical concentrations are to be measured the sensitivity can be improved by increasing τ, but the maximum value of τ is then usually set by the stability of the apparatus, the klystron and magnetic field in particular, rather than by the constancy of the radical concentration.

(ii) *Identification of the free radical species from the hyperfine structure*

It has been seen that the observed hyperfine pattern is very characteristic of the particular chemical groups included in the orbit of the unpaired electron. Radicals containing a group with n equally interacting protons will have a hyperfine pattern of $(n+1)$ lines with a Gaussian distribution of intensities. Overlapping spectra from different groups can often be separated, and the additional effects of other non-equivalent protons can be deduced, as in the case of dimesitylmethyl in section 5.5.2.

In a large number of biological studies, however, the radicals are rather complex and the hyperfine patterns are not well resolved.

Even in this case some general features can often be recognized and identified with the same spectrum in simpler compounds. The use of this ' analysis by comparison with simpler systems ' is illustrated by the X-irradiation results discussed later in section 9.6.

(*iii*)   *The study of the change in radical concentration with time or chemical treatment*

The study of change of radical concentration with time follows from the quantitative measurements under steady-state conditions. If the time scale is too fast for long sweep-rates and high sensitivity work, the radical system can often be deep frozen and the actual concentration caught at any desired time interval. If the interacting system is not destroyed in this process it can then be warmed, allowed to react for a further time, and then re-frozen. Alternatively, a series of different tubes containing the same interacting system can be frozen after different time intervals, and then studied afterwards at leisure. In this way rate constants of different chemical reactions or polymerization processes, and the like, can be determined.

(*iv*)   *The investigation of intermolecular electron exchange*

If electron transfer takes place between a free radical and non-radical molecules of the same compound, and the frequency of exchange is greater than the normal line width, as measured in frequency units, the observed resonance signal will be broadened by an amount proportional to the exchange frequency. The application of this fact to a study of electron transfer rates between aromatic negative ions has already been described in section 5.3.3. WARD and WEISSMAN[1] found that the mean lifetime of the naphthalene negative ion in the presence of 0·8M naphthalene molecules was 10^{-6} seconds, from which a rate constant of 10^6 litres mole^{-1} sec^{-1} at 30° C can be deduced. This same principle can be applied to a large number of other radical systems.

(*v*)   *The study of irradiation effects*

It has been seen in the preceding chapters that electron resonance is a very powerful tool in irradiation studies. It can, not only measure the concentration of the damage centres quantitatively, but also give information on the breakdown processes. Very interesting information can be obtained from the low energy irradiation studies employing u.v. or optical wavelengths, as well as from X- or γ-ray treatment. Both of these branches are of particular biological significance and they are considered in detail in sections 9.5 and 6.

(vi) *The investigation of free-radical mobilities*

Considerable information on the mobilities and internal motion of free-radical species can be obtained from a study of their stability under different trapping conditions. The work on radicals trapped in low temperature glasses is of particular interest in this respect since the viscosity of the trapping media can be varied very precisely in such cases. The mobilities and motion of the radicals can be studied not only by the decrease in signal strength, but also by the change in hyperfine patterns and line widths. Possible modes of radical reaction and decay can also be deduced from such experiments as summarized in Chapter 6.

It may not be possible to undertake each of the above investigations for any given system, of course, but the above headings indicate the broad applications for which electron resonance can be employed in free radical studies.

The particular importance of electron resonance investigations in biological and medical studies arises from the very general function that free radical reactions appear to have in metabolic and biochemical processes. The suggestion that many enzyme reactions involved chain reactions with free radicals as an intermediary was made nearly thirty years ago[2], and the classic work of MICHAELIS[3-5] and his co-workers on oxidation-reduction systems has strengthened this view very much of recent years. The existence of paramagnetic atoms in such enzymes as catalase and peroxidase also indicated that one-electron intermediates might play a very important part in normal biochemical processes. The advent of electron resonance has now provided a method whereby these various theories of free radical mechanism can be tested experimentally, and the actual radical concentration measured during any metabolic process or enzyme action.

Another field of particular biological and medical interest is that of irradiation studies. On the one hand the effect of optical irradiation on chloroplasts and its correlation with the general mechanism of photosynthesis is proving to be a very fascinating study. On the other hand electron resonance studies on X-irradiated biological material is giving information not only on the amount of irradiation damage, but also on the general types of free radical and molecular breakdown that are produced in different specimens.

The application of electron resonance to the biological and medical field is only just beginning, and a large amount of work will undoubtedly follow during the next few years. In this chapter a brief resumé is given of the work that has been carried out to date,

239

and an indication is added of the possible directions which future work will take. The results have been grouped together under five headings:

(i) The study of free radicals in living tissue.

(ii) The relation between free radicals and carcinogenic activity.

(iii) The study of oxidation-reduction systems and enzyme interaction.

(iv) Photosynthesis and optical absorption studies.

(v) X-irradiation of biological material.

9.2 The Study of Free Radicals in Living Tissue

The study of free radical concentration in living tissue is one of the most direct applications of electron resonance to biological problems, and an initial investigation was undertaken by Commoner, Townsend and Pake[6]. One of the great disadvantages of biological material in electron resonance studies is the presence of water in the specimen. Water, in its liquid phase, has a large non-resonant absorption due to the interaction of its large dipole moment with the microwave electric field. Hence aqueous samples of the normal size will cause large damping and a considerable reduction in Q factor and sensitivity if inserted into the cavity. This can be overcome by using much smaller samples, e.g. specimen tubes of 1 mm diameter, instead of the normal 5 mm diameter. Such a reduction in size of the sample automatically reduces the sensitivity, however, and hence the highest sensitivity of any spectrometer can never be achieved when aqueous solutions are being studied. It may be possible to obtain high sensitivity for aqueous studies by working in the high radiofrequency, instead of the microwave region. The loss due to the dipole interaction falls off with decreasing frequency, and if lumped circuit techniques are employed it is also possible to place the solution inside a coil, completely isolated from the radio-frequency electric field. The disadvantage of this is, of course, the fall in inherent sensitivity with decreasing frequency, due to the Boltzmann factor. It is possible that a compromise around the 200–500 Mc/s region may be achieved, however, with relatively high sensitivity and no dipolar damping by the water. It should be noted that this will only be of use for free-radical *concentration* studies, since one of the main factors increasing the sensitivity at the lower frequencies is the larger volume of sample that can be used.

If experiments are confined to the microwave region the loss due to the water can be eliminated either by freezing the specimen and performing the measurements at liquid nitrogen temperatures, or by freeze-drying the material before insertion in the cavity. In the latter method, however, care must be taken to ensure that no extra bonds are broken and radicals formed by the freeze-drying process. Both of these methods can, in fact, effect the radicals, and if sufficient concentration is present, studies with smaller specimen tubes at room temperature are to be preferred. COMMONER, TOWNSEND and PAKE[6] prepared their initial specimens by freeze-drying, and the powdered samples were inserted into the normal specimen holder of an X-band spectrometer. They made a survey of a large number of different tissues in this way and their initial results are summarized in *Table 9.1*. It is noticeable that the higher free-radical content is found in the metabolically active tissues such as green leaf, liver and kidney. It was also possible to show by fractionation experiments that the free radical concentration was associated with the protein components and that denaturation of the protein destroyed the radical concentration.

Table 9.1. Radical Concentration in Different Tissues

Material	Radical Concentration 10^{-8} mole/gm of dry weight
Nicotiana tabacum, leaf	65
Nicotiana tabacum, roots	10
Coleus, leaf	180
Barley, leaf	25
Digitalis, germinating seeds	10
Carrot, root	8
Beet, root	6
Rabbit, blood	25
Rabbit, muscle	20
Rabbit, brain	25
Rabbit, liver	60
Rabbit, lung	30
Rabbit, heart	35
Rabbit, kidney	55
Frog, eggs	200
Drosophilia, entire	4

* Taken from COMMONER, TOWNSEND and PAKE. *Nature*, Lond. 174 (1954) 689

Further confirmation that increased free radical concentration is associated with high metabolic activity was obtained from a study of ungerminated and germinating seeds, and from leaves subjected to varying amounts of illumination. Thus no measurable radical concentration was found in samples prepared from digitalis seeds before germination, but a concentration of 10^{-7} moles/gramme was obtained from seeds just after the emergence of the primary root.

The results obtained on the illuminated leaves are shown in *Figure 64*. The top curve is that obtained from leaves of barley seedlings grown in normal light. The second curve is that from leaves grown in the dark, and it is seen that the radical concentration is very much reduced. The second and third curves are for those grown in the dark and then irradiated for 6 and 12 hours respectively by 100 ft.-candles of illumination from fluorescent lamps. It is seen that this treatment raises the radical concentration to its normal level.

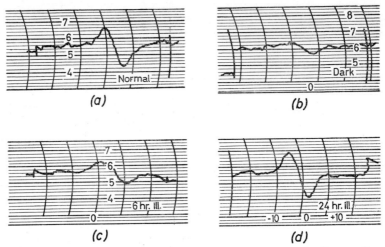

Figure 64. Free radical concentration in illuminated barley leaves. (a) *From seedlings grown in normal illumination;* (b) *From seedlings grown in the dark;* (c) *Grown in dark and then illuminated for 6 hours with 100 ft.-candles;* (d) *After illumination for 24 hours with 100 ft.-candles (After Commoner, Townsend and Pake[6])*

These preliminary measurements therefore support the theories of Michaelis[5] that metabolic electron transfer induces the formation of a small steady-state concentration of free radicals. Moreover these radical concentrations are characterized by the fact that they appear to be associated with proteins, that they attain concentrations which vary with the level of metabolic activity and that they are of magnitudes which are consistent with the estimations[7] of total electron carrier content of the tissues.

COMMONER, TOWNSEND and PAKE[6] also found that there was a high free radical concentration in melanin, which is a pigmentation found in various biological tissues. It has been shown[8] that melanin formation may be induced in animals by u.v. or ionizing radiation, and this is further evidence for the production of free radicals during irradiation of living tissue.

9.3 FREE RADICALS AND CARCINOGENIC ACTIVITY

Free radicals have been postulated as taking part in carcinogenic activity by several workers[9-15] and LIPKIN *et al.*[12] have suggested that the carcinogenic activity of certain large ring structures may be related to their ability to form negative ion free radicals with mild reducing agents. In contrast to this, non-carcinogenic hydrocarbons, such as naphthalene, require very strong reducing agents before such radicals are formed. The technique of electron resonance now provides a powerful new tool by which these theories may be tested.

The most direct experiment on the subject of carcinogenic activity is to measure the free-radical concentration in growing tumour material. Measurements of the radical concentration of normal healthy tissue and of cancerous tissue from the same part of the body have been made in the author's laboratory and elsewhere[16], but no direct correlation between cancerous tissue and higher radical concentration has yet been found. This does not mean, however, that the *initiation* of abnormal growth is not by a free-radical mechanism. To check on this point experiments are now being carried out on the radical concentrations present in various carcinogenic compounds themselves.

In this connection, experiments have recently been performed on the radical concentration present in cigarette smoke[17]. The cigarettes were fixed in a small manifold which was connected to a filter pump so that intermittent drawing was possible, as illustrated in *Figure 65*. A sealed-off length of $\frac{1}{4}$ inch diameter tubing led down from this manifold, as shown, and was immersed in liquid oxygen during the smoking. In this way some of the smoke was condensed into the tube and any active radicals present would remain unaltered when in such a deep-frozen condition. When about $\frac{1}{2}$ g. of condensate had collected, the apparatus was transferred to the electron resonance spectrometer and the tube was inserted into an H_{014} cavity surrounded by liquid oxygen. The radical concentration existing under these conditions was then measured. The tube was then removed and warmed to $60°$ C for about five minutes so that any active radicals could recombine. The tube and contents were then deep-frozen, re-inserted in the spectrometer and the radical concentration again measured. In this way it was possible to distinguish between short-lived, active radicals and those that were resonance or sterically stabilized.

The concentration of radicals obtained in the frozen smoke condensate before warming was of the order of 10^{15} free electrons per gramme. It should be noted that this condensate contains large amounts of solid carbon dioxide and ice. On warming, the condensate separated into an aqueous and an organic or tarry

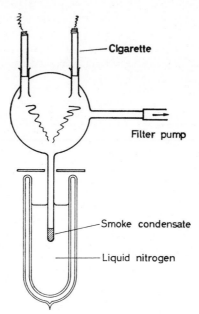

Figure 65. Apparatus for trapping cigarette smoke at low temperatures (After Lyons, Gibson and Ingram[17])

phase. No radical concentration at all could be detected in the aqueous phase, and that in the organic phase was reduced by a factor of approximately six. The radicals that could still be detected in the latter were found to be highly stabilized and no diminution of their concentration was found after several days. It would therefore appear from these experiments that the tar constituents of cigarette smoke contain about 6×10^{15} free electrons/gramme. Some of these radicals, being relatively short-lived, disappear when the condensate is warmed to 60° C for some minutes. The remainder, however, appear to be very stable and long-lived. It is probable that these stable radicals are very similar to those formed by pyrolysis of other organic matter as discussed in Chapter 7.

These experiments therefore show that there is a relatively high concentration of both active and stabilized free-radicals in cigarette smoke when it is first formed, and it is possible that either or both of these might act as carcinogenic agents. It is important to note in this connection that most bioassays in tobacco carcinogenesis have been carried out using relatively old condensates, and it is now evident that these will be deficient in the unstable free-radicals initially present in the smoke.

The concentration of stable free radicals in atmospheric soot is about one hundred times larger than in cigarette smoke, but, as with the polycyclic hydrocarbons[18], adsorption on comparatively larger particles and further stabilization as a result, is likely to render them inaccessible to the cells. It is evident that a large amount of systematic work will have to be performed before the correlation between free radical concentration and carcinogenic activity is established, but these initial experiments show that electron resonance should be of considerable help in these studies.

9.4 OXIDATION-REDUCTION AND ENZYME SYSTEMS

It has already been seen that free radicals have been postulated as necessary intermediates in biological oxidation-reduction systems[5], and that the preliminary measurements of COMMONER, TOWNSEND and PAKE[6] appear to support this hypothesis. Free radical processes have also been postulated as taking part in most enzyme reactions[19], and kinetic studies often show that chain processes are present. A large number of electron resonance studies of enzyme reactions are now in progress to investigate the details of these reaction mechanisms. At the time of writing, no detailed results on actual enzyme systems have been published, but some preliminary basic work on simpler oxidation-reduction systems has been reported.

One series of organic compounds closely related to biologically important molecules are the phthalocyanines. These are large planar molecules consisting of a conjugated ring system very similar to that of the porphyrin ring, and containing either two protons or a divalent metal atom at the centre. If these are oxidized by such agents as ceric sulphate a two-stage oxidation process takes place via a transient intermediate stage which often gives the solution a markedly different colour for a few seconds. There has been considerable speculation[20] as to whether this intermediate oxidation stage involved a change of valency of the central metal atom or the

R   245

liberation of an unpaired electron in the ring system. Electron resonance studies are able to give a very direct answer to such a problem as this, since different valence states of the central metal atom will have characteristic g-values well displaced from the free-spin value, whereas a mobile electron in the ring system will have a narrow ' free-radical ' resonance line very close to $g=2{\cdot}0$.

It was found experimentally[21] that all the phthalocyanines studied had intermediate oxidation states that gave a narrow resonance line with a g-value very close to that of a free spin. This was therefore conclusive proof that the oxidation process involved mobile electrons in the conjugated ring system. In some cases[21] it was possible to follow the oxidation process through its different stages and watch the growth and decay of the intermediate on the oscilloscope of the resonance spectrometer.

Another oxidation system of considerable biological importance is the oxidation of ferrihaemoglobin to its metastable state. This system has been studied in some detail[22,23] by electron resonance. The systematic analysis of these results is a good illustration of the fact that it is important to consider all possible interactions when investigating biological systems. As in the case of the phthalocyanines, the oxidation intermediate will either involve an unpaired electron in the conjugated ring system or a change in the valency state of the central iron atom. Chemical investigations[24,25] give inconclusive evidence as to which of these mechanisms is actually present.

In the initial electron resonance studies[22] it was found that a strong narrow ' free-radical ' line with a g-value of $2{\cdot}003$ was obtained when methaemoglobin or metmyoglobin was oxidized by hydrogen peroxide or periodate. By analogy with the results obtained from the phthalocyanines it was therefore assumed that the oxidation involved electron removal from the π-orbitals of the porphyrin ring. Further more detailed, and systematic investigations of this system[23] confirmed that there is a close connection between the free radical and the peroxide compound, as shown by four separate sets of results. First, the formation of the free radical is specific for the methaemoglobin peroxide reaction, since it does not occur in model systems such as serum albumin–Fe–H_2O_2 or metmyoglobin cyanide–Fe–H_2O_2. Secondly, removal of excess peroxide by catalase does not affect the free radical. Thirdly, the amount of free radical is proportional to the initial concentration of metmyoglobin, under comparable conditions. Fourthly, the free radical is destroyed by substances reducing the peroxide compound (ferrocyanide, nitrite, iodide, p-cresol).

In spite of the above conclusions, however, there are further results which show that the free radical cannot be identified with the peroxide compound itself. The reasons for this can be summarized as follows. First, the calculated concentration of free radical in the most favourable case was only 9 per cent of the peroxide compound. Secondly, its concentration varied widely under conditions in which the amount of peroxide compound remained approximately constant. Thus increase of pH from 6 to 10 decreased the free radical concentration by 80 per cent, while the peroxide compound was formed equally well at these pH values. Thirdly, preaddition of stoichiometric quantities of ferrocyanide to the metmyoglobin before formation of peroxide compound almost eliminated the free radical. Such treatment has been found[24] to remove the first oxidizing equivalent of hydrogen peroxide lost during the formation of the peroxide compound.

These experiments therefore suggest that the free radical is not the peroxide compound itself, but a product of oxidation by the first oxidizing equivalent, which is likely to be a hydroxyl radical[24]. The free radical signal is thus probably due to oxidation of part of the globin molecule, and although it takes place at the same time as the peroxide compound is formed it is not associated directly with this oxidation mechanism. It is nevertheless of considerable interest as it is an example of a reversible univalent oxidation state in a biological system which does not involve valence changes of a transition metal. The fact that the formation of the actual peroxide compound involved a change in the binding of the iron atom was demonstrated conclusively by observing the metmyoglobin resonance at $g=6\cdot0$[26,27] as oxidation occurred. This resonance was found to decrease as the peroxide compound was formed indicating a change from ionic to covalent binding[28] as the oxidation took place.

It may be noted in this connection that a detailed study of the different derivatives of haemoglobin by electron resonance not only allows the actual orbitals involved in the binding to be determined[28], but also gives detailed structural information on the porphyrin and histidine planes[27-31].

This work on the oxidation states of haemoglobin and myoglobin indicates the necessity for careful systematic studies if the electron resonance spectra of these biological systems is to be correctly interpreted. This remark will apply with even greater force to the catalase and peroxidase[23], and other complex enzyme systems that are being currently investigated.

247

It may be noted that a kinetic study of these enzyme reactions not only allows the variation of free radical concentration to be followed, but also enables correlations to be made between this and the concentration of any paramagnetic atoms that are present. The exact role of different metallic ions in enzyme reactions has been a matter of speculation for some time, and in some cases they are assumed to play an essential step in the oxidation-reduction mechanism, and in others they appear to enter only as impurities. If the concentration of the metallic ions is monitored[48] at the same time as that of the free radicals these effects can be clearly distinguished. As an example of this, measurements have recently been made[47] on dehydrogenase acting on an octanyl substrate, and the concentration of the cupric ions was found to follow approximately the same kinetics as the free radical. Care must again be taken in interpreting these results in detail, since more than one free radical may be present, but electron resonance will be able to give some very important information in such studies as these.

9.5 PHOTOSYNTHESIS AND OPTICAL ABSORPTION STUDIES

It has already been seen in section 9.2 and *Figure 64* that the free-radical concentration present in leaves is dependent on the amount of illumination that they have received. This in itself strongly suggests that free radicals play an important role in photosynthesis, and it had been proposed previously that the excitation of chlorophyll during photosynthesis involved mono-[5] or bi-radical[32] formation. In the initial experiments[6] the leaves were lyophilized before insertion into the cavity, and hence kinetic studies on the specimens were not possible. In a more recent series of experiments, however, COMMONER, HEISE and TOWNSEND[33] were able to study aqueous suspensions of chloroplasts, *in situ* in the cavity, under different conditions of illumination. This was achieved by the use of very small specimen tubes ($\frac{1}{2}$ mm × 1 mm cross-section) in conjunction with a high-sensitivity 100 kc/s field modulation spectrometer

The chloroplasts, which are known to be responsible for the essential steps of photosynthesis[32], were prepared from leaves of *nicotiana tabacum* by gentle hand maceration in an ice bath in a buffer solution of pH 8. After filtration and centrifugation, the chloroplasts were resuspended in 55 per cent sucrose, and stored in the cold. The spectrometer employed a reflection-type cavity, and had a 2 mm diameter hole in the shorting end through which the light from a car headlamp was focussed. In this way the free-radical concentration in the chloroplast suspension could be studied under different conditions of illumination and the rate of growth and decay measured accurately.

It was found that there was only a very small radical concentration in the absence of any illumination, but that this increased by about sixfold as soon as the 50 c.p. headlamp was switched on. The absorption line had an appearance very similar to those of *Figure 64*, with a *g*-value of $2 \cdot 002 \pm 0 \cdot 001$ and a width of 10 gauss, as observed for the free-radicals in other metabolically active tissue. Rapid tracing of the signal enabled an accurate plot of the growth of radical concentration to be obtained. It was found that the concentration rose exponentially to a steady value after onset of illumination with an exponential time constant of 12 seconds. The decay of radical concentration when the lamp was switched off was also of an exponential form, but with a somewhat longer constant of 45 seconds. These results show conclusively that a steady-state radical concentration was being measured and not a system of trapped or stabilized radicals.

The variation of radical concentration with intensity and the wavelength of the incident illumination was also studied[33]. The concentration was found to rise with increasing intensity but reach a saturation value at high levels, and this is also true of the photosynthetic activity of both chloroplasts and whole cells. It was also shown that the free radical concentration was produced by the same wavelength range, 4,000 to 7,000 Å, which is responsible for photosynthesis. These experiments therefore provide conclusive evidence that one-electron intermediates with unstable molecular configurations are produced during photosynthetic reactions.

Further measurements on illuminated chloroplasts were then made by Sogo, Pon and Calvin[34]. In particular they studied the variation of radical rise and decay time with temperature of the specimen. Their results are summarized in *Table 9.2* and the most striking feature of these results is that the signal growth time is approximately the same when the chloroplasts are frozen at -140° C as when they are at room temperature. This fact would appear to rule out the ordinary enzymatic oxidation-reduction reactions as the free-radical intermediates in photosynthesis. The longer decay time at -140° C also suggests that excitation to the triplet state is not responsible for the observed signal, and, as seen in the last chapter, it is unlikely that this would be observed in any case. It would therefore seem that the intermediate associated with photosynthesis is some form of electron produced by a dissociated bond or trapped in a quasi-crystalline lattice. It is noticeable[34] in this connection that the lines at room temperature show evidence of exchange narrowing, but are wider at the lower temperatures indicating a reduced mobility of the unpaired electrons.

Table 9.2.*   *Growth and Decay Time of Radical Concentration in Photosynthetic Material*

Substance	Light Intensity quanta/sec.	Temp. °C	Signal Growth Time	Signal Decay Time
Dried leaves	10^{15}	25	minutes	hours
Dried whole chloroplasts	10^{15}	{ 25	minutes	hours
		60	seconds	seconds
Wet whole chloroplast	10^{15}	{ 25	seconds	1 minute
		–140	seconds	hours
Wet small chloroplast fragments	10^{15}	25	seconds	minutes
Wet large chloroplast fragments	{ 10^{15}	24	30 seconds	30 seconds
	10^{16}	25	6 seconds	30 seconds
	10^{16}	–140	10 seconds	hours

* Taken from Sogo, Pon and Calvin. *Proc. Nat. acad. Sci.* 43 (1957) 387

Tollin and Calvin[35] have also made detailed studies of the luminescence of similar chloroplast samples and have correlated these results with the electron resonance measurements. As a result of this it would appear that the hypothesis of electron trapping and the production of 'holes' is the most plausible theory of photosynthetic reaction. This semi-conductor theory of chloroplast action has been proposed by several investigators[36–38] and is also supported by glow-curve and resistance measurements[38]. On this theory the luminescence which occurs immediately after illumination ceases will be largely due to radiative recombination of nearly free electrons and holes. Following this, thermal excitation of electrons and holes from the shallowest traps will produce further emission and the decay constant will be a function of the trap depth and the temperature. The close correlation of the luminescence and electron resonance is explained in this way, and also the much longer decay times obtained at low temperatures, when the thermal excitation energy is much smaller while the trap depth remains more or less the same.

These initial electron resonance studies of photosynthesis are only of a preliminary nature, however, and the interpretation of the results must still be considered as somewhat speculative. They do illustrate very well, however, the kind of information that can be obtained, and the technique should prove a very powerful complement to all the normal luminescence and phosphorescence studies.

9.6   X-Irradiation of Biological Material

A study of the breakdown processes which occur in living tissue as a result of X- or γ-ray irradiation is one of the most important problems in modern medical physics. Electron resonance is one of the most direct and sensitive methods of investigation in this particular field, and the initial results have already shown that a large variety of different spectra can be obtained from different specimens. It is probably wise to commence this section with a word of caution, however, since it is very easy to draw false conclusions when employing high energies to irradiate complex biological material. The energy of the incident quanta is not only great enough to break a very large number of different bonds, but the secondary radical and non-radical species can also produce further change in the cell structure. Thus most biological specimens contain a large percentage of water, and it is known that the OH˙ radicals formed in this by primary photolysis will combine and produce relatively high concentrations of hydrogen peroxide. This is very toxic for a large number of cellular reactions, and may easily produce breakdown processes, the products of which can then be further attacked by primary or secondary radicals. Deductions of the mode of breakdown from the observed electron resonance spectra should therefore be made very tentatively until all possible mechanisms have been considered.

Electron resonance studies of this kind can be divided into two broad categories:

(*i*) quantitative studies of free radical production as a function of radiation dosage, and

(*ii*) qualitative analysis of the types of radical structure formed in the breakdown process.

In quantitative studies it is possible to study either the dynamic concentrations which are formed *in situ* in the resonator, or the larger concentrations obtained in a solid or viscous medium by a suitable trapping technique. In order to obtain dynamic concentrations of sufficient intensity it is necessary to employ very powerful radiation sources, and this often requires pulse techniques. The use of pulsed irradiation can also give useful additional information. Thus it is possible, in principle, to trigger the electron resonance spectrometer at a specified time delay after the irradiation pulse, and hence obtain a series of measurements on the free radical decay between successive pulses. High sensitivity detection with long-time constants is not feasible under such conditions, however, and this is another reason why high radiation dosages must be employed.

Several laboratories are working on systems suitable for these quantitative dynamic studies, but to date most information has come from the qualitative analysis of the different radical structures present in the breakdown process.

The identification of different radical species from the hyperfine pattern of their electron resonance spectra has been considered at some length in Chapter 6. It was seen there that it is often possible to pick out the presence of specific groups in the presence of other

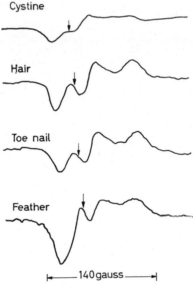

Figure 66. Spectra from X-irradiated cystine and natural proteins. The arrow indicates the position of g=2·0023 (After Gordy, Ard and Shields[39])

free radicals, and this method of analysis can now be extended to the more complex biological compounds. The results of GORDY, ARD and SHIELDS[39] on X-irradiated cystine and various fibrous proteins may be taken as a specific example of this. The electron resonance spectra obtained from cystine, hair, nail, and feather are shown in *Figure 66*, and they are seen to be very similar. This similarity is even more striking when it is found that the spectra are not symmetrically placed around g=2, and have splittings which vary with the strength of the external field. It would therefore appear that in each case the free radicals must have an unpaired electron strongly localized on a sulphur or oxygen atom with an associated anisotropic g-factor. It is noticeable in this connection that the spectra are very similar to those obtained from frozen solutions of sulphur in oleum[40].

Gordy, Ard and Shields[39] in fact explain this spectra as due to an unpaired electron localized on the S–S bond of the cystine. Thus if an electron is ejected by the irradiation from some point in the molecule, the vacancy probably moves to the S atoms to form an additional lower energy three-electron bond, as represented by the structural formula:

$$\left(\begin{array}{c} \text{HO} \quad \text{H} \quad \text{H} \qquad\qquad \text{H} \quad \text{H} \quad \text{OH} \\ \text{C} - \text{C} - \text{C} - \text{S} \cdots \text{S} - \text{C} - \text{C} - \text{C} \\ \text{O} \quad \text{H}_2\text{N} \quad \text{H} \qquad \text{H} \quad \text{NH}_2 \quad \text{OH} \end{array} \right)^{+}$$

There is thus now a hole shared by the two sulphur atoms through exchange of their electrons, and the loss of an electron from the sulphur lone pair orbitals tends to strengthen rather than weaken the bond in this case.

From the marked similarity of the spectra of the irradiated hair, feather and toe-nail it would appear that the radicals in these proteins must also be associated with their cystine content. The above hypothesis also explains why only the cystine spectrum is observed although the cystine group is a small fraction of the total amino-acid content of the protein. If the three electron S—S bond represents a lower energy state the cystine is likely to donate an unpaired electron to any other ionized group formed by the irradiation. In this way the two sulphur atoms of the cystine molecule act as an electron reservoir and supply electrons to fill vacancies at other points in the protein molecule, and such a mechanism should very much reduce the damage produced by the irradiation. It is therefore evident that in these particular cases the electron resonance spectra have not only shown that the cystine type grouping is present in these proteins, but have also indicated that the breakdown process is transferred to the S–S bond, and molecular fracture is therefore not so likely to take place in these compounds.

These results indicate the type of reasoning that is used when deducing radiation effects from resonance spectra of complex molecules. Another example is given by the spectra observed from irradiated glycylglycine, silk, cattle hide and fish scale[39]. A symmetrical doublet with a frequency-independent splitting of 12 gauss is obtained in all of these cases. This is attributed to the direct dipole-dipole interaction between the hydrogen bonding proton and the unpaired electron localized on an oxygen atom of the adjacent polypeptide chain. It is noticeable in this connection that X-ray investigation shows that silk has an elongated polypeptide chain structure with the adjacent chains linked by hydrogen bonds between the $C=O$ and NH groups.

This method of analysis by ' correlation of similar spectra ' will have to be employed in most of the biological studies in the immediate future. The conclusions deduced in this way must be considered as tentative and often as just one possible explanation among several others. As further results are obtained, however, a greater background of information will be built up on the most likely type of hyperfine interaction present in any particular system. The detailed interpretation of the observed spectra will thus be facilitated, and the conclusions will be more definite and reliable.

GORDY et al.[41-45] have started a series of measurements on the electron resonance spectra observed from various X-irradiated proteins, to try and build up some systematic data for this purpose. With this in mind the simpler peptide and polypeptides were studied first[41], so that their spectra could then be compared with those obtained from the more complex proteins. These investigations included X-irradiated glycylglycylglycine, DL-alanyl-DL-alanine, acetyl-DL-alanine, glycyl-DL-valine, acetyl-DL-leucine, and DL-alanylglycylglycine. The electron resonance spectra of the more complex proteins such as histone, insulin, haemoglobin, and albumin were then investigated[42], and compared with those of the simpler proteins. It was found that two types of pattern were obtained in each case. One of these is very similar to that obtained from the irradiated cystine, and the other is a doublet similar to that obtained from the irradiated glycylglycine. It would therefore appear that the electron donating power of the S–S bonds, and the hydrogen bonding across the polypeptide chains may be general features in most protein structures. The work is also being extended to irradiated hormones and vitamins[43] such as progesterine parathyroid, vitamin A, vitamin K, ascorbic acid, and also to irradiated nucleic acids[44] such as DNA, RNA, adenosine, cytidine and inosine. These measurements are all of a preliminary nature at the moment, and much careful and systematic work will have to be done before any definite conclusions can be established. The field is a very wide one, however, and when sufficient systematic measurements have been made some detailed information on the different mechanisms associated with irradiation damage should emerge.

9.7 THE TREND OF FUTURE WORK

It will be seen from the preceding sections that the application of electron resonance to systems of biological and medical interest has only just started. In nearly every case the interpretation of the results is hampered by the very complex nature of the material under investigation. It will therefore be some time before a sufficient number of correlated spectra are obtained, and detailed conclusions can be drawn. One exception to this is the use of electron resonance for quantitative analysis of trace paramagnetic elements, such as manganese or copper, as they occur in natural tissue[45,46]. In this case an unequivocal answer can be obtained directly without any destruction of the material under investigation.

The role of free radicals in biochemical processes is of such importance, however, that a large amount of work is likely to be carried out in this field during the next few years, and electron resonance is an ideal tool for such studies. It would appear that the three fields of greatest interest will be

(i) the study of enzyme reactions

(ii) the investigation of free-radical concentration and carcino-genic activity, and

(iii) the study of irradiation effects.

The initial studies on the intermediates present in oxidation-reduction systems have already given strong support for Michaelis' theory[5] on enzyme reaction. Further investigation of different enzyme systems, and in particular of the growth and decay of radical concentration with the concentration of other constituents present, should give some very interesting data on the role actually played by the enzyme interaction. In a similar way, a systematic study of the free-radical concentration present in different carcino-genic compounds should produce conclusive evidence on the role of free radicals in the initiation of cancerous growth. The X- and γ-irradiation studies will probably be of a wider and more general character, and quantitative data will not only become available on the amount of damage produced by any given radiation dosage, but it should also be possible, in due course, to elucidate the mechanism of breakdown and suggest means whereby this can be prevented or modified.

It should be added in conclusion that, although these three lines of research appear to be those which are most rapidly developing at the moment, some new field of much greater interest and impor-tance may easily open up at any time. One of the most fascinating features of electron resonance is the way in which it is being applied to subjects of ever-widening variety and background.

REFERENCES

1 WARD, R. L. and WEISSMAN, S. I. *J. Amer. chem. Soc.* 76 (1954) 3612
2 HABER, F. and WILLSTÄTTER, R. *Berichte.* 64 (1931) 2844
3 MICHAELIS, L., BOEKER, G. F. and REBER, R. K. *J. Amer. chem. Soc.* 60 (1938) 202
4 — SCHUBERT, M. P., REBER, R. K., KUCK, J. A. and GRANICK, S. *Ibid.* 60 (1938) 1678
5 — *The Enzymes.* Academic Press, 1951, pp. 1–46
6 COMMONER, B., TOWNSEND, J. and PAKE, G. E. *Nature, Lond.* 174 (1954) 689
7 McILWAIN, H. *Ibid.* 158 (1946) 898
8 BLUM, H. F. ' Biology of Melanomas '. *Acad. Sci., N.Y.* 4 (1948) 388
9 KENSLER, C. J., DEXTER, S. O. and RHOADS, C. P. *Cancer Res.* 2 (1942) 1
10 PARK, H. F. *J. Phys. & Coll. Chem.* 54 (1950) 1383
11 FITZHUGH, A. F. *Science,* 118 (1953) 783
12 LIPKIN, PAUL, TOWNSEND, J. and WEISSMAN, S. I. *Ibid.* 117 (1953) 534
13 BRUES, A. M. and BARRON, E. S. G. *Ann. Rev. Biochem.* 20 (1951) 350
14 OPPENHEIMER, B. S., OPPENHEIMER, E. T., STOUT, A. D., DANISHEFSKY, I. and EIRCH, F. R. *Science,* 118 (1953) 783 and *Cancer Res.* 15 (1955) 333
15 NASH, T. *Nature, Lond.* 179 (1957) 868
16 PAKE, G. E. *Disc. Faraday Soc.* 19 (1955) 181
17 LYONS, M. J., GIBSON, J. F. and INGRAM, D. J. E. *Nature, Lond.* 182 (1958) 1003
18 STEINER, P. E. *Cancer Res.* 14 (1954) 103
19 WATERS, W. A. *The Chemistry of Free Radicals.* Oxford University Press, 1946, Ch. 12
20 CAHILL, A. E. and TAUBE, H. *J. Amer. chem. Soc.* 73 (1951) 2847
21 GEORGE, P., INGRAM, D. J. E. and BENNETT, J. *Ibid.* 79 (1957) 1870
22 GIBSON, J. F. and INGRAM, D. J. E. *Nature, Lond.* 178 (1956) 871
23 — — and NICHOLLS, P. *Ibid.* 182 (1958) 1398
24 GEORGE, P. and IRVINE, D. *Biochem. J.* 52 (1952) 511
25 — — *Ibid.* 58 (1954) 188; 60 (1955) 596
26 BENNETT, J. E. and INGRAM, D. J. E. *Nature, Lond.* 177 (1956) 275
27 INGRAM, D. J. E., GIBSON, J. F. and PERUTZ, M. F. *Ibid.* 178 (1956) 906
28 GIBSON, J. F., INGRAM, D. J. E. and GRIFFITH, J. S. *Ibid.* 180 (1957) 29
29 INGRAM, D. J. E. and KENDREW, J. C. *Ibid.* 178 (1958) 905
30 BENNETT, J. E., GIBSON, J. F. and INGRAM, D. J. E. *Proc. roy. Soc. A* 240 (1957) 67
31 — — — HAUGHTON, T. M., KERKUT, G. and MUNDAY, K. *Phys. Med. & Biol.* 1 (1957) 309
32 HILL, R. and WITTINGHAM, C. P. *Photosynthesis.* Methuen, 1955
33 COMMONER, B., HEISE, J. J. and TOWNSEND, J. *Proc. Nat. acad. Sci.* 42 (1956) 710
34 SOGO, P. B., PON, N. G. and CALVIN, M. *Ibid.* 43 (1957) 387
35 TOLLIN, G. and CALVIN, M. *Ibid.* 43 (1957) 895
36 KATZ, E. *Photosynthesis in Plants.* Iowa State College Press, 1949, p. 291
37 BRADLEY, D. F. and CALVIN, M. *Proc. Nat. acad. Sci.* 41 (1955) 563
38 ARNOLD, W. and SHERWOOD, H. K. *Ibid.* 43 (1957) 105
39 GORDY, W., ARD, W. B. and SHIELDS, H. *Ibid.* 41 (1955) 983
40 INGRAM, D. J. E. and SYMONS, M. C. R. *J. chem. Soc.* (1957) 2437
41 McCORMICK, G. and GORDY, W. *Bull. Amer. phys. Soc.* 1 (1956) 200
42 GORDY, W. and SHIELDS, H. *Ibid.* 1 (1956) 267
43 REXROAD, H. N. and GORDY, W. *Ibid.* 1 (1956) 200
44 SHIELDS, H. and GORDY, W. *Ibid.* 1 (1956) 267
45 — ARD, W. B. and GORDY, W. *Nature, Lond.* 177 (1956) 984
46 INGRAM, D. J. E. and BENNETT, J. E. *J. chem. Phys.* 22 (1954) 1136
47 BEINERT, H. E. Quoted in *Nature, Lond.* 181 (1958) XXXV
48 COHN, M. and TOWNSEND, J. *Nature, Lond.* 173 (1954) 1090
MALMSTRÖM, G., VÄNGÅRD, T. and LARSSON, M. *Ibid.* 183 (1958)

APPENDIX

ELECTRON RESONANCE EQUIPMENT AVAILABLE COMMERCIALLY IN 1958.

THESE are not intended to be complete lists of all equipment available, but summarize those at present in use with which the author is acquainted.

I. COMPLETE ELECTRON RESONANCE SPECTROMETERS

    (i) *Electronique Medicale et Industrielle, 17, Rue Montbrun, Paris 14, France.*
        A simple crystal-video reflection cavity spectrometer, with magnetic field sweep at 50 c/s and oscilloscope presentation.
        Approximate price (1958) £1,000.

    (ii) *Microwave Instruments Ltd., West Chirton Industrial Estate, North Shields, Northumberland, England.*
        X-band spectrometer employing 100 kc/s magnetic field modulation, with coil inside the microwave cavity. Klystron is locked to cavity by means of a waveguide discriminator system. (Magnets with 4 in. or 8 in. diameter pole faces are being developed for use with this spectrometer.)
        Approximate price (1958) £3,000.

    (iii) *Strand Laboratories Inc., 294 Centre Street, Newton 58, Mass., U.S.A.*
        Two types of spectrometer available for X-band operation, one locked to the sample cavity, the other to external reference cavity. 6 kc/s magnetic field modulation is employed, together with ferrite circulator in the microwave bridge. Cavities can operate from 4°K to above room temperature and will take samples up to $\frac{1}{2}$ in. diameter.
        Approximate price (1959) £2,000.

    (iv) *Varian Associates, 611, Hansen Way, Palo Alto, California, U.S.A.*
        X-band spectrometer employing audio frequency magnetic field modulation and phase sensitive detection. Obtains high sensitivity by very good stabilization

of klystron and magnet. (Varian magnets suitable for this spectrometer are listed below.)
Approximate price (1958) £5,000.

II. ELECTROMAGNETS FOR E.S.R.

A. FOR HIGH RESOLUTION WORK

(i) *Metropolitan-Vickers Electrical Co., Trafford Park, Manchester, England.*

9 in. diameter pole face, provides fields of 15,000 gauss with 1·5 in, gap. Has high resistance water-cooled windings with power consumption of 2 kw. Can be supplied with matching power supply and also flux stabiliser.
Approximate price (1958) £6,300 (with power supply).

(ii) *Newport Instruments (Scientific and Mobile) Ltd., Newport Pagnell, Buckinghamshire, England.*

Type D. 8 in. diameter pole face, provides fields of 12,000 gauss in a 2·25 in. gap. Has low resistance windings and can be supplied with matching power supply.
Approximate price (1958) £2,700 (with power supply).
Type E. 7 in. diameter coned to 6 in. diameter pole face, provides fields of 4,200 gauss in 2·25 in. gap. Has low resistance windings and can be supplied with matching power supply.
Approximate price (1958) £1,200 (with power supply).

(iii) *Varian Associates, 611, Hansen Way, Palo Alto, California, U.S.A.*

(*a*) Standard 12 in. diameter pole face provides fields to 13,000 gauss with 1·75 in. gap, but other gap geometries are available. Has high resistance water-cooled windings, and can be supplied with matching power supply.
Approximate price (1958) £5,500 (with power supply).
(*b*) Standard 6 in. diameter pole face provides fields to 9,000 gauss with 1·75 in. gap, but other gap geometries available. Has high resistance water-cooled windings, and can be supplied with matching power supply.
Approximate price (1958) £3,600 (with power supply).

B. For Low Resolution Work

(i) *Metropolitan Vickers-Electrical Co., Trafford Park, Manchester, England.*
6 in. diameter pole face provides fields to 3,500 gauss with 1·5 in. gap. Has high resistance windings with power consumption of 175 watts. Can be supplied with matching power supply.
Approximate price (1958) £640 (with power supply).

(ii) *Newport Instruments (Scientific and Mobile) Ltd., Newport Pagnell, Buckinghamshire, England.*
Type A. 4 in. diameter pole face provides fields to 6,000 gauss with 1·5 in. gap. Has low resistance windings and can be supplied with matching power supply.
Approximate price (1958) Magnet £300. Power Supply £600.

C. Other Magnets

The following British firms are also developing magnets for Electron Spin Resonance:—
Fairey Aviation Co., Heston Aerodrome, Hounslow, Middlesex.
Mullards Ltd., Component Division, Mullard House, Torrington Place, London, W.C.1.

III. x-band klystrons. Can be supplied by:

E.E.V. Co. Ltd., Chelmsford, England.
E.M.I. Ltd., Hayes, Middlesex, England.
Mullard Ltd., Torrington Place, London, W.C.1.
Raytheon Mfg. Co., Waltham 54, Massachusetts, U.S.A.
Varian Associates Ltd., Palo Alto, California, U.S.A.
Western Electric Co., 120, Broadway, New York, U.S.A.

IV. x-band crystals.

B.T.H. Ltd., Rugby, England.
G.E.C. Ltd., Magnet House, Kingsway, London, W.C.2, England.
Sylvania Electric Co., 1221 West 3rd St., Williamsport, U.S.A.
Western Electric Co., 120, Broadway, New York. U.S.A.

V. X-BAND BOLOMETERS.

Narda Corporation., 66, Main Street, Mineola, New York, U.S.A.
Sperry Gyroscope Co., Great Neck, New York, U.S.A.

VI. WAVEGUIDE COMPONENTS. (English Firms only.)

Elliott Bros. Ltd., Century Works, London, S.E.13.
Microwave Instruments Ltd., West Chirton Industrial Estate, North Shields, Northumberland.
Mid-Century Microwave Gear Ltd., 3, Oakwood Place, Stanley Road, West Croydon, Surrey.
Sanders (Electronics) Ltd., 48, Dover Street, London, W.1.

VII. KLYSTRON POWER SUPPLIES. (English Firms only.)

Airmec Ltd., High Wycombe, Buckinghamshire.
Microwave Instruments Ltd., West Chirton Industrial Estate, North Shields, Northumberland.
Solartron Electronic Ltd., Thames Ditton, Surrey.

VIII. FREE RADICAL MARKER, HYDRAZYL.

Fluka A.G. Chemische Fabrik, Buchs. S.G., Switzerland.

AUTHOR INDEX

263

264

SUBJECT INDEX

Page numbers in heavy type indicate main references to the subject.

267

269

271